T0361798

Water Management for Large Cities

Efficient and equitable water, wastewater and stormwater management for the megacities is becoming an increasingly complex task. This book focuses on water management in its totality for megacities, including their technical, social, economic, legal, institutional and environmental dimensions through a series of specially invited case studies from different megacities of the world.

At present, around one in two of the earth's 6.3 billion people live in urban areas. Each year, the world population grows by around 80 million. Practically all of this growth is urban, primarily due to migration. The world's urban population is expected to reach 5 billion by 2030, which is nearly two-thirds more than in 2000, and would mean that 60% of world's population will live in urban areas.

The case studies analysed include some of the most interesting and challenging megacities of this planet, Dhaka, Istanbul, Jakarta, Johannesburg, Mexico City, ArRiyadh and São Paulo. They assess different aspects of how water is intermingled in the overall development milieu. *Water Management for Large Cities* considers the magnitudes, nature and extent of the present and future challenges and how these could be met in socially acceptable and cost-effective ways. The international contributors are all acknowledged water experts.

This book was previously published as a special issue of the *International Journal of Water Resources Development*.

Cecilia Tortajada is the Vice President of the Third World Centre for Water Management, Mexico

Olli Varis is a Senior Researcher at the Helsinki University of Technology

Jan Lundqvist is a Professor at the Swedish International Water Institute

Asit K. Biswas is President of the Third World Centre for Water Management, Mexico

Water Management for Large Cities

Edited by Cecilia Tortajada, Olli Varis,
Jan Lundqvist and Asit K. Biswas

LONDON AND NEW YORK

First published 2006 by Routledge
2 Park Square, Milton Park, Abingdon, Oxon, OX14 4RN, UK

Simultaneously published in the USA and Canada
by Routledge
270 Madison Ave, New York, NY 10016

Routledge is an imprint of the Taylor & Francis Group, an informa business

© 2006 Taylor & Francis Ltd

Typeset in Times 10/12pt by the Alden Group Oxford
Printed and bound in Great Britain by Antony Rowe Ltd, Chippenham, Wiltshire

British Library Cataloguing in Publication Data
A catalogue record for this book is available from the British Library

Library of Congress Cataloging in Publication Data

ISBN10 0–415–41363–x
ISBN13 978-0-415–41363–3

CONTENTS

v

Preface

ASIT K. BISWAS

Since around 1960, the speed, scale, scope and complexity of urban transformation processes of the developing world have become a formidable task to manage. It is now undoubtedly one of the most important challenges facing humankind. The urban population of the world never exceeded 7% until about 1800. The whole rural/urban dynamics was completely changed, especially in the developed word, by the Industrial Revolution. At present, nearly 75% of the inhabitants in the Western world live in urban areas.

With advances in medical and health services, and as life expectancies in the developing world started to increase during the post-1950 period, the urbanization process in developing countries accelerated and gained momentum. However, there is one fundamental difference between the urbanization processes witnessed by developed and developing countries. The urban growth rates in developed countries were gradual, and thus there was time to plan and manage this growth. The economies of these countries were also growing as they were undergoing urbanization. This ensured that financial resources were available to manage this incremental growth over a long period of time.

In contrast, the urban growth in developing countries has been very rapid. For example, the first megacity of the world, New York, had 150 years to plan and manage its urban population which increased by 8 million. In contrast, Mexico City witnessed doubling of the urban influx that New York had faced (16 million) but it had to manage this growth in only one-third of the time (50 years) New York had. Furthermore, Mexico City neither had the resources nor management capacity to manage this rapid urbanization.

With increasing globalization, acceleration of information and communication revolution and increasing chasms between the incomes and lifestyles of urban and rural population, the next two decades are likely to witness major transformations in the urban centres of the developing world, and perhaps some urban–rural conflicts over development priorities.

China is a good example of the urbanization problems the developing world may face. At present, 40% of Chinese workers are engaged in the agricultural sector but they account for only about 15% of the country's economic output. Unless the lifestyle of the rural Chinese can be improved significantly within about a decade or so, many Chinese from the rural areas are likely to migrate to urban areas looking for employment in the industrial and service sectors. In fact, the OECD has estimated that between 70 to 100 million rural workers are likely to migrate to urban areas between 2000 and 2010, looking for a better standard of living. If so, such massive rural–urban migration in about one decade in only one country will be unprecedented in human history.

Managing this rural exodus to urban areas in Asian countries like China and India, where urbanization has not been as advanced as in the Latin American region, will be a major challenge in the coming years. Provision of basic services like water, sanitation, education, health, housing, transportation, etc., will put accelerating pressure on the policy-makers to formulate and implement new and innovative policies which can ensure that poor people have a reasonably decent standard of living.

In order to objectively and comprehensively discuss water management problems of major urban centres of the developing world, the Third World Centre for Water Management, the Water Resources Laboratory of the Helsinki University of Technology and Stockholm International Water Institute organized a seminar in Stockholm. All the papers for this volume, except the one on Singapore, were specially commissioned from leading water experts from the selected regions. These papers were then discussed in depth during this Seminar, and the authors then revised their papers in the light of the comments received at Stockholm. The paper on Singapore is a much-expanded version of the one that was prepared for the 2006 Human Development Report of the United Nations Development Programme. Together, all these contributions provide an overall picture of the current situations and future prospects for water management as a whole for major urban centres of the developing world.

I would like to take this opportunity to express our appreciation to the Water Resources Laboratory of the Helsinki University of Technology for providing intellectual and financial support to this activity, and to Stockholm International Water Institute for the technical and organization support that we received for the Seminar.

Dr Olli Varis of Helsinki University of Technology, Professor Jan Lundqvist on behalf of the Stockholm International Water Institute, and Dr Cecilia Tortajada from the Third World Centre for Water Management, were instrumental for the organization of the Seminar, and the subsequent editing of the papers. We are most grateful to these three internationally well-known water personalities for their support which made both the Seminar and this volume possible.

Water Management for Major Urban Centres

ASIT K. BISWAS

Introduction

From the dawn of human history, water has been an essential requirement for the survival of humans and ecosystems. Water has played a critical role in human development. For much of history, droughts and floods have periodically inflicted serious damage to society.

As the urban centres started to develop in a very serious manner, especially after the Industrial Revolution in the developed world, the provision of clean water and disposal of wastewater and stormwater became increasingly important issues. However, by the early part of the second half of the 20th century, these problems for much of the Western World had been solved. In addition, the Western economies had become stronger and more resilient than ever before, and thus floods and droughts had progressively less and less impact on such societies. When they occurred, they could either be controlled, or their adverse impacts could be reduced by the construction of the appropriate infrastructure and increasingly more efficient management practices.

However, the problems have mostly worsened for developing countries during the second half of the 20th century. Increasing population growth, rapid urbanization, inadequate levels of economic development and the absence of appropriate management and technical capacities meant that an increasing number of the urban population did not have access to basic services such as water, sanitation and stormwater management. At the global level, this issue received attention in 1976, during the United Nations Conference on Human Settlement, held in Vancouver, Canada. Subsequently, it was addressed firmly and directly by the United Nations Water Conference, held in Mar del Plata, in 1977. The Water Conference recommended that the period 1981–90 be declared as the International Water Supply and Sanitation Decade. The objective of the Decade was to provide clean water and sanitation to every person in the world by 1990.

A retrospective analysis indicates that the Decade had a major impact by putting water supply and sanitation higher up the international political agenda, in both developed and developing countries, than otherwise might have been the case. It also contributed to accelerated progress in terms of provision of clean water and sanitation to a large number of people, both in actual and percentage terms. However, in spite of this noteworthy progress, the goals for the Decade proved to be too ambitious to reach. In the end, the world fell far short of achieving its objectives.

The challenge was subsequently partially picked up by the Millennium Development Goals (MDGs), one of which explicitly stipulated that, by 2015, the objective will be to "reduce by half the proportion of people without sustainable access to safe drinking water".

This MDG differs in three important aspects with reference to the Decade objectives. First, its objective is more modest: reducing by half the proportion of people who do not have access to a water supply compared to the universal coverage championed by the Decade. Second, the MDG, unlike the Decade goals, did not specifically refer to sanitation. This, of course, is somewhat surprising, since any water supply project introduces new water in the urban areas, nearly all of which subsequently becomes wastewater, which must be collected, treated and disposed of in an environmentally-sound manner. Provision of urban water supply, even if it becomes universal by 2015, will not be sustainable by itself, unless adequate arrangements can be made for wastewater collection, treatment and disposal. Third, the MDG has 50% more time to achieve its objective compared to the Decade: 15 years against a decade. However, current analyses indicate that even though one-third of the time period to achieve this MDG goal is over, pro rata progress has been much slower than necessary to achieve the final objective.

The sanitation issue was later considered and agreed to at the Johannesburg Summit in 2002. Accordingly, if the MDG goal for water supply and the Johannesburg recommendation on sanitation are to be achieved by 2015, accelerated progress has to be made during the next 10 years. Such an accelerated progress will require that the best practices in providing access to water supply and sanitation from different parts of the world be carefully analysed and assessed. These best cases could then be modified appropriately for possible application in other developing countries, in line with their social, economic, political, legal and institutional requirements. Regrettably, some 25 years after the Decade started, serious, comprehensive and objective evaluations of the best examples from different parts of the world are mostly still missing at present.

Urbanization and Water Management

Any discussion of urbanization should begin with a definition of what is 'urban'. Unfortunately, there is no international agreement as to what is an urban area. It is almost universally agreed that any settlement having more than 20 000 people is urban. However, many countries consider areas of less than 20 000 people as urban as well. Thus, aggregating global statistics on urban areas and analysing them becomes problematical. For example, if the Indian government considered settlements of more than 5000 people to be urban, as some countries do, India will have a predominantly urban population.

Countries generally define 'urban' based on one or more criteria, which include population size, population density, social and economic factors such as percentage of people involved in non-agricultural activities, administrative or political status of the settlement or census designations. The UN World Urbanization Prospects, 1996 Revision, points out that 46% of the countries it considered defined urban on the basis of administrative criteria, 22% used population numbers and sometimes population densities, 17% used other criteria, 10% had no definition and 4% considered their countries as entirely urban or entirely rural.

Historically, human societies had a predominantly rural lifestyle. For example, in 1800, only about 3% of the global population lived in urban areas. This increased to 14% by 1900, when 12 cities had more than 1 million people. The urbanization process advanced dramatically during the next 50 years. Thus, by 1950, the world had become almost 30% urbanized, and the number of cities with more than 1 million inhabitants had increased to 83.

During the period 1950–2000, the growth rates of the urban population in almost all countries was higher than the rural population. However, the world is heterogeneous, and there were significant differences in growth rates in different urban centres. If the growth rates of the 524 urban centres that had more than 750 000 population in 2000 are considered, 41 of them had growth rates of less than 1% and six had negative growth rates during the period 1950–75. This trend of low levels of growth rates continued during 1975–2000, when 122 cities had growth rates of less than 1% per year, and 21 of them had witnessed negative growth rates. Most of these cities were in developed countries and China. This trend is expected to accelerate during 2000–15. Table 1 shows some of these trends.

General historical experience has been that as the population of a city increases, after a certain point its population growth rates start to decline. Accordingly, cities that tend to grow at higher rates generally have smaller populations. For example, during 2000–15, of all the major urban centres, only Dhaka and Lagos will have annual growth rates of around 4.0%. Eight other megacities will have growth rates of less than 1%.

However, there are exceptions. For example, the population of Mexico City was 2.9 million and that of Sao Paulo 2.5 million in 1950. Even with such large populations, their annual growth rates were 5.2% and 5.6% respectively. Consequently, their

Table 1. Population growth rates for cities with more than 750 000 inhabitants in 2000

Population growth rates	Number of cities		
	1950–75	1975–2000	2000–15
More than 5%	28	3	0
More than 5%	130	51	6

Source: UN Population Division (2002).

Table 2. Population growth rates (%) of megacities in the sequence of their evolution, 1950–2015

City	Growth rate (%)		
	1950–75	1975–2000	2000–15
New York	4.2	1.2	0.2
Tokyo	1.0	0.2	0.5
Shanghai	3.1	0.5	0.4
Mexico City	5.2	2.1	0.8
Sao Paulo	5.6	2.2	1.1
Mumbai	3.6	3.1	2.3
Los Angeles	3.2	1.6	0.6
Kolkata	2.3	2.0	1.7
Dhaka	6.6	7.0	4.0
Delhi	4.6	4.1	3.5
Buenos Aires	2.4	1.1	0.6
Jakarta	4.8	3.3	3.0
Osaka	3.5	0.4	0.0
Beijing	3.1	1.0	0.5
Rio de Janeiro	4.0	1.2	0.5
Karachi	5.4	3.7	3.2

Source: UN Population Division (2002).

population increased by a factor of four by 1975, making them the first two megacities of Latin America having more than 10 million inhabitants.

Similarly, during the period 1975–2000, Dhaka started to grow at an annual rate of 7%, as a result of which its population increased by six times by 2000, when it became a megacity of more than 10 million people. The city with the highest growth rate during this period was ArRiyadh, which increased annually by 7.4%. On the basis of current trends, no large city is expected to grow at such a high rate during the 2000–15 period (Table 2).

Initially, the United Nations defined a megacity as an urban area having more than 8 million people. As urban centres became bigger and bigger, a megacity was later redefined as having more than 10 million people.

If the current definition is accepted, the world had only one megacitiy in 1950, New York. By 1975, the number of megacities had increased to five, and, by 2000, to 16. It is estimated that by 2015, this number will increase to 21, and New York will become only the 7th largest city in the world (Table 3). In 2000, only four megacities out of 16, were from the developed world. By 2015, even though the total number of megacities is expected to increase to 21, not even a single new megacity will join this list from developed countries.

According to the analyses carried out by the United Nations Population Division (2001 Revision), the following trends can be discerned between 2000 and 2030 in terms of urbanization:

• The global urban population will increase from 2.9 billion in 2000 to 5.0 billion in 2030. Nearly all of this increase will take place in developing countries. The urban population is expected to grow at an annual average rate of 1.9%, compared to global population growth rate of 1%.
• The urban population of developed countries will increase marginally from 0.9 billion in 2000 to 1.0 billion in 2030.

Table 3. Evolution of megacities, 1950–2015 (population in millions)

1950		1975		2000		2015 (projected)	
City	Population	City	Population	City	Population	City	Population
New York	12 339	Tokyo	19 771	Tokyo	26 444	Tokyo	27 190
		New York	15 880	Mexico City	18 066	Dhaka	22 766
		Shanghai	11 443	Sao Paulo	17 962	Mumbai	22 577
		Mexico City	10 691	New York	16 732	Sao Paulo	21 229
		Sao Paulo	10 333	Mumbai	16 086	Delhi	20 884
				Los Angeles	13 213	Mexico City	20 434
				Kolkata	13 058	New York	17 944
				Shanghai	12 887	Jakarta	17 268
				Dhaka	12 519	Kolkata	16 747
				Delhi	12 441	Karachi	16 197
				Buenos Aires	12 024	Lagos	15 966
				Jakarta	11 018	Los Angeles	14 494
				Osaka	11 013	Shanghai	13 598
				Beijing	10 839	Buenos Aires	13 185
				Rio de Janeiro	10 652	Metro Manila	12 579
				Karachi	10 032	Beijing	11 671
						Rio de Janeiro	11 543
						Cairo	11 531
						Istanbul	11 362
						Osaka	11 013
						Tianjin	10 319

Source: UN Population Division (2002).

Table 4. Distribution of urban population, 1975–2015

Types of urban settlements in terms of population	World		Developing countries	
	Annual average population increase (millions)			
	1975–2000	2000–15	1975–2000	2000–15
More than 10 million	6.3	7.7	5.0	7.5
5–10 million	1.9	6.3	2.8	5.9
1–5 million	13.7	19.1	10.9	17.3
0.5–1 million	4.5	4.3	4.2	4.5
Less than 0.5 million	26.3	29.8	23.2	28.2
Total	52.8	67.2	46.3	63.4

Source: UN Population Division (2002).

- The rural population of developing countries is likely to increase by 0.2% annually during this period. The rural growth rate is expected to turn negative in 2025 for the first time in human history. Asia will continue to have the largest rural population in the world during 2000–30, with 2.297 billion in 2000 and declining to 2.271 billion by 2030. In contrast, Africa's rural population will increase from 498 million in 2000 to 702 million in 2030. The rural population of Europe, Latin America and the Caribbean and North America will also decline during this period.
- Globally, urban cities with populations of between 5 to 10 million will expand the fastest during 2000–15, and cities with below 500 000 inhabitants will account for the highest percentage of such populations (Table 4). The trends are likely to be similar for developing countries.

While the megacities of the developing world have attracted the most attention from the various international organizations in recent years in terms of provision of adequate water supply and sanitation services, it should be noted that they account for a very small percentage of the global population, even though they consume a lion's share of the national resources and interest in terms of the various necessary infrastructure development and management. If megacities are defined as those having more than 10 million residents, only 3.7% of the global population lived in such large urban agglomerations in 2000. By 2015, this is likely to increase to 4.7%. The percentage of people living in cities of between 5 to 10 million are even less than in megacities: 2.8% in 2000, increasing to 3.7% in 2015. In other words, percentages of the global population living in cities of more than 5 million will increase from 6.5 in 2000 to 8.4 in 2015.

While the megacities present tremendous management challenges at present, and will continue to do so in the foreseeable future, much of the recent urban growths are being witnessed in medium to small sized urban centres of developing countries. This trend is likely to continue in the coming decades, and may even accelerate. Thus, in 2000 cities with less than 500 000 inhabitants accounted for 24.8% of the global population (nearly seven times that of megacities) and this is expected to rise to 27% in 2015.

It should be noted that urbanization and the formation of major urban metropolises are not new phenomena. For example, cities such as London or New York started to grow significantly in the 19th century. However, two major differences should be noted which have made the urbanization process and provision of water supply and sanitation services in the megacities of the developed world very different in comparison to their counterparts in developing countries, nearly one century later.

First, is the rate of growth. The development of these earlier urban centres in the developed world was a gradual process. For example, most of the population growth in cities such as London and New York was spread over nearly a century. Gradual growth rates enabled these cities to progressively and effectively develop the necessary infrastructure and the capacities to manage their water supply and sewerage services. Since it was a gradual development, the increases in population were thus manageable.

In contrast, most of the urbanization in the large cities of the developing world such as Dhaka, Mexico City, Sao Paulo or Jakarta occurred during the post-1950 period, and the really explosive growth generally took place after 1960. These major urban centres simply could not cope with the very high and continually increasing urbanization. They were not only unprepared to manage such explosive growths, but also they did not have the financial and management capacities to manage this work. Thus, the overall quality of life declined rapidly during such periods of high urbanization. As noted earlier, between 1950 and 1975, the population of Mexico City increased more than fourfold, a significantly larger increase compared to what the urban centres of the developed world had witnessed earlier.

To a certain extent, many of these megacities could handle the provision of a water supply, but they generally fell progressively behind in constructing and managing sewage and wastewater treatment facilities. Even in a region like Latin America, where many cities made reasonably good progress in installing sewerage systems, concomitant progress did not occur in wastewater treatments. Currently, for the most part, less than 10% of collected sewage in major urban centres of Latin America are treated properly and then disposed of in an environmentally-safe way. Thus, in major Latin American cities, ranging from Bogota to Buenos Aires, and Mexico City to Santiago, millions of cubic metres of untreated, or partially treated, sewage is discharged daily into nearby water bodies. Many of the governments often claim that some 20 to 30% of their wastewaters are treated, but these are highly inflated figures that are not compatible with the real situations.

Second, as the urban centres of the industrialized countries expanded, their economies were improving concomitantly as well. Accordingly, these centres were economically able to harness financial resources to provide its citizens with appropriate water supply and sewerage services. For a country like Japan, it could invest heavily in the construction of urban infrastructure, including water supply, sewerage and flood control services after the Second World War because its economy continued to expand very significantly during the post-1950 period. Such extensive infrastructure development and major improvements in management practices in Japan meant that water losses due to leakages from the urban water supply systems could be reduced drastically from an immediate post-war estimate of 90% to about 8%, which is one of the lowest losses encountered anywhere in the world at present. Equally, cities like Tokyo spent enormous amount of funds during the post-1950 period to control urban flooding which would not have been possible if Japan's economy had not expanded during this period as well.

In stark contrast to the above, during the past four decades, economies of the developing world have not performed very well. Issues such as high public debts, poor governance and inefficient resource allocations have ensured that the investments needed to construct all types of new urban water and sanitation-related infrastructure and maintain the existing ones have not been forthcoming. Lack of proper planning, poor management and practices and pervasive corruption have further aggravated the situation in many urban centres.

While considerable progress has been made in recent years in providing drinking water in urban areas, commensurate advances in sanitation have been, for the most part, missing in much of the developing world. In the past, sanitation has proved to be a poor cousin to water supply. Thus, it is not surprising that the Millennium Development Goals considered water supply but not sanitation. Consequently, the major water problem that developing countries are likely to face in the coming decades is not likely to be physical water scarcity, although it will not be an easy problem to solve, but continued deterioration of water quality. Water sources within and near major urban centres of the developing world, from Dhaka to Mexico City, are already heavily contaminated. In the absence of adequate water quality management practices and the absence of political will, the local situations are deteriorating steadily. Herein is likely to be the future water crisis of the developing world.

While continuing urbanization poses a major challenge in terms of provision of water supply and sanitation services, the importance and the contributions of such urban centres towards the development of stronger and more stable national economies should not be underestimated. It has been estimated that the urban areas of the developing world, which contained about 47.2% of the total population in 2000, contribute nearly two-thirds of their total Gross National Products, and also play an equally important prominent role in terms of social development and cultural enhancement. Accordingly, the urbanization process presents both challenges as well as opportunities. The main issue is how to manage the urbanization process properly for all its inhabitants.

A main problem for the major urban centres thus stems from the fact that the rates of urbanization have generally far exceeded the capacities of the national and the local governments to soundly plan and manage the demographic transition processes efficiently, equitably and sustainably. Provision and maintenance of the needed infrastructural development, services and employment are critical. The accelerating urban growth rates have generally overwhelmed the limited management capacities and resources of the governments at all levels. Unquestionably, unplanned and poorly managed urbanization processes have been an important source of social and environmental stress in all developing countries. The impacts of this poorly managed process have manifested in extensive air, water, land and noise pollution, which have, and will continue to have, major impacts on human health and welfare of the urban dwellers of the developing world for many years to come. The problem has been further compounded by increasingly skewed income distribution which is continuing to worsen with time, high rates of unemployment and under-employment, corruption at all levels, and high crime rates.

The two major problems faced by the major urban centres have further intensified an already difficult and complex situation. First is the sudden fast rate of vertical growth, often after decades, or even centuries, of primarily horizontal expansion, especially in the central business areas. This contributed to a sudden surge in population densities of these areas, with concomitant high water requirements and generation high waste loads per unit area. The existing water supply and sanitation services and the poor planning and management capabilities of the concerned authorities have mostly been unable to cope successfully with such almost instantaneous growths in higher demands in water and sanitation services in such areas.

Second, the overall water-related problem of the large urban centre is further compounded by the presence of informal and squatter settlements. Such settlements may account for 30–60% of the total urban population. For example, it is estimated that approximately half of Mumbai's population lives in such squatter areas. These areas are highly congested, leaving very little, or no space, for the provision of an in-house water supply and/or public sanitation facilities. As more and more poor people from rural areas migrate to urban centres, areas and densities of such squatter settlements often increase regularly.

The situation is further compounded by the fact that nearly all levels of governments have generally given a lower priority to informal settlements in terms of developments. Areas where rich and important people live receive higher priority in terms of budgets. In addition, urban planners often believe that adequate cost recovery for the provision of services to such settlements are not possible, since they are inhabited by poor, or very poor, people and/or the cities do not have adequate resources to provide highly subsidized water and sanitation services to these settlements on a long-term, reliable basis. Accordingly, informal settlements are often neglected, or receive lower priority in terms of management time, allocation of resources, and thus services, compared to rich and middle-class areas. Furthermore, the population of these informal settlements often grows steadily due to the continuing influx of rural migrants, searching for better economic conditions, and thus quality of life. Accordingly, whatever limited services are available in the informal settlements become overwhelmed with the arrival of the new migrants. The limited water supply and sanitation services that may be available become progressively less and less adequate for serving an ever-increasing population. This contributes to progressive reduction of services available that were inadequate to start with, and this deterioration, in turn, further increases the environmental and health conditions of the people living in such areas.

There are some signs that these situations have started to change. The work of activist NGOs are bringing the plights of the poor people in the squatter settlements to the attention of national and international organizations and the media. This has increased the awareness of the problem, and has started to improve the overall water and sanitation conditions in informal settlements in many developing countries. In addition, private sector concessions in several urban centres now specifically stipulate performance indicators, which they must meet in terms of access to water supply and sanitation in such areas, as well as stormwater disposal. All these new developments have started to improve the existing situations, but much remains to be done.

Constraints to Urban Water Availability

The provision of clean drinking water to the rapidly growing urban centres of the developing world and safe disposal of wastewater faces numerous constraints, which are complex and interrelated. Only a very few resolutions or declarations of various international fora (the United Nations General Assembly resolution on the International Water Supply and Sanitation Decade is an exception), have had any visible and perceptible impacts. Even for the Decade, it was evident during the 1970s that its goals, however laudable, would be impossible to achieve unless major structural changes were made in terms of resource allocation to the sector, both nationally and internationally. Not surprisingly, not only were the goals not met, but also fell considerably short. However, the Decade did manage to put water supply and sanitation issues higher up in the political agendas of many governments of developing countries and external support agencies, especially towards the beginning and end of this period.

Unfortunately, no serious evaluation was carried out as to what were the actual impacts of the Decade *per se*. Accordingly, the lessons that could have been learnt, both positive and negative, from the Decade experiences are unknown at present. For example, would the global situation have been different had there been no Decade? If so, how would it have been different, by how much, and why? What were the regional variations in terms of achieving the goals of the Decade, and why? Anecdotal evidence indicates that developing countries concerned and many external support agencies were already becoming aware of the importance of water supply and sanitation issues by the early 1970s, and that part of the developments that have occurred since 1980 probably would have occurred even without the Decade. However, the Decade provided a strong focus for water supply and sanitation issues, and some countries took advantage of this fact to accelerate their programme of activities in this area. Thus, overall, the Decade probably contributed to an improvement in access to water supply and sanitation services, perhaps significantly, which otherwise may not have occurred.

There are numerous major constraints that have to be overcome simultaneously before full access to water supply and sanitation services can be assured in major urban centres of the world. Overcoming these constraints will not be an easy task, nor is it likely to be achieved universally within the next decade, irrespective of the Millennium Development Goals. However, the future is not completely bleak, and there are many positive and encouraging signs. For example, in the coming decades, as the population growth rates in developing countries continue to decline, gradually they will have to run less and less fast to stay at the same place. As various developing countries approach a stationary population, they are likely to be in somewhat better positions to be able to provide clean water and sanitation services to all their citizens on a reliable and sustainable basis. Population stabilization will mean that they will not have to continually chase a moving target, as is the case at present. However, these developments are likely to be noticeable in about one generation, and not in the immediate future.

Some of the major constraints to achieve the ambitious goal of providing clean water and sewage services to the large urban centres of the world will be briefly discussed herein.

Water Scarcity

Water scarcity presents both a challenge and an opportunity in terms of urban water supply and sanitation. It is a challenge because new sources of water which could be developed cost-effectively for major urban areas of the developing world, are mostly not available. From Istanbul to Johannesburg, and Jakarta to Mexico City, there are simply no new sources of water that could be harnessed economically and in a socially and environmentally-acceptable manner which can quench the continually increasing urban-industrial thirst.

Since the existing sources of water that could be developed cost-effectively have already been developed, or are in the process of development, and water that has been harnessed has already been fully allocated (in fact, in many cases over-allocated), an additional supply of drinking water can only be obtained by transferring water which is currently being used by other sectors, especially agriculture. National policies, explicitly or implicitly, give the highest priority to the domestic sector amongst all the uses. However, socially and politically, it would not be an easy task to transfer water from the agricultural to the domestic sector. Some such transfers can already be noted. However, these transfers generally did not occur because of deliberate policy decisions: they occurred as an indirect result of other policy decisions.

A good example is the case of water availability in Chennai (formerly Madras) in India. In order to improve water use efficiencies in the agricultural sector, it was decided to line the irrigation canals. A direct result of this policy was a *de facto* transfer of water from the agricultural to the urban sector. Farmers who used the seepage water from the unlined canals to grow crops suddenly discovered their livelihood was gone. Most of them were forced to migrate to the city to seek alternative ways of survival, and thus adding to the squatter population of Chennai. Such practices are exacerbating the conflicts between agricultural and urban water users around many major cities all over the developing world, ranging from Manila to Mexico City. These conflicts are likely to increase significantly in the future, both in terms of their intensities as well as their geographical distributions, as various interest groups clamour for more and more water and supply becomes restricted.

Urbanization also brings opportunities. Urban centres may be important users of water, but such use takes place within a limited geographical area. Since domestic use does not contribute to actual consumption of water, all the water that is being supplied to the households can be recaptured as wastewater through sewage networks. If this wastewater can then be properly treated, it could then serve as a 'new' source of water, as is currently the case for Singapore. While treated wastewater may be restricted to specific types of water uses due to quality considerations and cultural reasons, it can be used for agricultural, industrial and commercial purposes, thereby releasing higher quality water for those uses that warrant it. Furthermore, the marginal cost of providing additional good quality water of the same volume as the treated wastewater would generally be much higher, and the time required to obtain a similar quantity of additional good quality water from a new source would be much longer.

Wastewater is produced in urban areas, irrespective of whether it is used or not. Equally, it is essential that wastewater be treated adequately in order to reduce environmental and health hazards for the people living in and around the urban areas. At present existing unsatisfactory wastewater disposal practices can be observed in the urban areas of most developing countries.

Thus, increasing water scarcity could at least in one sense be considered to be an opportunity that could encourage urban areas to collect and treat wastewater properly so that it could be subsequently used as a 'new' source to alleviate water scarcity.

High Economic Costs

Economic factors are becoming an increasingly important consideration for the provision of a water supply and sanitation to the urban areas of developing countries.

For much of the developing world, for the most part, all the easily exploitable sources of water have already been developed, or are currently in the process of development. This means that the water sources that are yet to be developed are geographically, technologically and environmentally, more complex to handle. Accordingly, the costs of harnessing and bringing this new water to the urban areas are becoming very high in real terms, especially compared to the cost of the earlier, or even the present, generation of water projects. For example, the average cost of providing storage for each cubic metre of river flow in Japan has increased nearly fourfold during the past 15 years. Approximately 20–30% of this additional cost can be attributed to the new social and environmental requirements which were not considered earlier. The major part of the additional cost is due to the fact that the new projects are inherently more complex to construct techno-economically, and are often located in more inhospitable terrains. Therefore, the construction costs of these new projects are significantly higher when compared with already completed projects, or those under construction.

The situation is not much better either with the treatment of wastewater. Most of the wastewaters produced in the urban areas of developing countries are either not treated at all, or receive inadequate treatment. Many governments, ranging form Egypt to Mexico, have often legislated high water quality standards because of internal and/or external political reasons, and faulty and incomplete appreciation of the problem. No consideration is generally given to the fact that the standards that are appropriate and can be implemented for cities such as London or New York may be irrelevant, impossible to implement and often may even be counterproductive for Lima or Yaounde. Equally, no serious analyses are generally carried out as to whether the standards adopted are essential for health reasons, or whether the countries concerned have the necessary financial resources, management capabilities and legal enforcement capacities to implement the stipulated standards. Not surprisingly, promulgation of such inappropriate standards generally have not even helped in maintaining, let alone improving, the quality of effluent discharged to water bodies in and around urban centres. Proper water quality management will undoubtedly be one of the major water problems of the future.

Financing and Management Constraints

Availability of adequate funds and release of the funds in a timely manner to operate and maintain existing water and wastewater facilities in the urban centres of the developing world is a major constraint at present. Water utilities in developing countries are predominantly in the public sector, although private sector involvement is being considered in one form or another in some parts of the world. The operation and maintenance of existing water supply and wastewater treatment systems, as well as the construction of new systems, are often constrained by lack of funds.

The economic situation is further compounded by inadequate pricing and inefficient billing and bill collection systems in most utilities of major urban centres. While this situation has improved very significantly in some countries like Sri Lanka (Biswas *et al.*, 2005) and Morocco, the situation has remained very similar in many major cities of Asia, Africa and Latin America. A review of the Asian urban cities indicates the following (ADB, 2003 and personal observations) shortcomings:

- Less than 50% of the connections are metered properly. Currently, major cities such as Kolkata have no metering, and many cities have very little metering. Regular monitoring and replacement of faulty meters are exceptions rather than norms. In practical terms, metering has become irrelevant since the cost of reading and maintaining meters, and billing in a city is often significantly higher than the total amount collected from the consumers.
- Monthly household water bills in many major cities such as Beijing, Tianjin, Hanoi, Mumbai and Tashkent are less than $1.00. In contrast, the average monthly bills for well-managed utilities such as those in Hong Kong or Singapore are significantly higher. Very low monthly bills encourage extravagant

consumption and high wastage rates. Electricity to water bill ratios average 18.5 for Faisalabad, 12.7 for Karachi, 9.2 for Tashkent, 7.8 for Kathmandu and 7.7 for Delhi. These ratios, whenever they are over 4, generally indicate low water tariffs and poor management practices.

- The financial management of many utilities leaves much to be desired. For example, accounts receivable should be less than the equivalent of 3 months of sale. However, it is nearly 20 months for Mumbai, 17 months for Karachi and around 11 months for Dhaka and Shanghai.
- The utilities have different concepts of what constitutes operation and management expenses. Many normal operation and maintenance expenditures are often left for rectification by new investment projects. Such expenditures include replacement of pipes, valves, water meters, service vehicles and reduction of unaccounted waters. Thus, major investments are made in constructing new systems, which are subsequently not properly maintained. This steadily increases system inefficiencies due to continuing deterioration. Accordingly, new investment projects have to be conceived to rehabilitate the badly managed systems. The process contributes not only to inefficient use of capital but also the system efficiencies start to decline steadily from the inception because of poor operation and maintenance practices. During the entire process, the customers of the water utilities receive a poor and unsatisfactory service, consistent with low water and sanitation charges.
- Utilities are often overstaffed, and for the most part the staff available are not properly trained and are inefficiently used. This generally contributes to inefficiency and low financial return. For example, staff per 1000 connections is around 2 for a well-managed utility like Singapore, but very high for Tianjin (49.9), Mumbai (33.3), Beijing (27.2) and Chennai (25.9). The high ratios indicate poor efficiency and management practices. Low staff ratios could, in a few cases, indicate that many services are being contracted out to the private sector.

Management Constraints

One of the major reasons for the poor performance and/or efficiencies of water utilities is their poor management. The main reasons for the poor management stems from two factors: unattractive salaries and regular political interference in management practices and decision-making processes of the water utilities.

In many urban areas, the management remuneration rates are determined by the government salaries, since the utilities are in the public sector, and thus follow public service rules. Since private sector salaries are much higher than their public sector counterparts, bright and competent managers generally tend to gravitate towards private sector enterprises, where, in addition, there is also much less day-to-day interference from the politicians, and thus better job satisfaction rates. It must be appreciated by the politicians that multi-million dollar water and wastewater utilities cannot be managed efficiently by unqualified and inexperienced managers, with continual political interference, ranging from the recruitment of staff to how resources are allocated.

Analyses of compensation packages of water managers in the Asian developing countries show very wide variances. For some utilities, the annual salaries are less than $1000, but in a few countries it could be as high as $100 000. Not surprisingly, the well-managed urban utilities of Hong Kong, Singapore, Taipei and Kuala Lumpur pay high salaries, and consequently they tend to attract and retain good managers and accountants.

The more efficient utilities also give their managers more financial autonomy and authority to make prompt and efficient management decisions. For example, the Metropolitan Waterworks Authority of Bangkok has the financial autonomy to raise investment funds in the local bond market. Its overall performance is good, and hence the general public subscribes to its bonds. Similarly, the Public Utilities Board of Singapore has considerable autonomy in staffing, finance and procurement of goods and services. It also has a clearly enunciated tariff policy, which although high, is acceptable to the people and politicians as a whole. Thus, its policies can be implemented efficiently, without any political interference.

In contrast, the situation in Mexico is very different. The head of a water utility is a political patronage position. The head is appointed primarily because of his/her political connections to the party and the existing political structure, and not because of professional and management expertise (Rodriguez & O'Neal, 2006). The entire top management structure of the water utility changes with each new Mayor, thus preventing the formulation and implementation of any long-term coherent policy and plans. The average stay of a utility manager in Mexico is only two years. The absence of a firewall between electoral politics and utility management invariably contributes to serious management deficiencies.

Another issue is the extensive use of public taps in certain major urban centres. It is a good indicator of poor management practices. The water utilities that are better managed in Asia, such as those in Bangkok, Kuala Lumpur or Singapore, do not have public taps, because they already have 100% coverage. Public taps often indicate lower levels of service, as well as higher water wastages. In addition, utilities cannot recover revenue from such taps, and city authorities are reluctant to subsidize them directly from city taxes.

Environmental and Health Issues

Water quality and health are major factors that must be considered for efficient urban water and wastewater management.

In a vast majority of the urban centres of the developing world, drinking water directly from city supply systems entails considerable health risks. Thus, not surprisingly, people often boil the tap waters prior to drinking. In addition, sales of water filters and bottled water have increased, and continue to increase in recent years. For example, the sale of bottled water in all urban centres of the developing world has increased exponentially during the past decade. This explosion in demand for bottled water is now a common phenomenon in all developing countries, ranging from Brazil to India and Dhaka to Mexico. This demand stems from only one factor, that is, the intense fear of the consumers, often justified, that the water supplied by the public utilities is not of good quality, and thus drinking it may entail significant health risks.

Ironically, even though consumers have taken to drinking bottled water in ever increasing quantities, the quality of water bottled often leaves much to be desired. Absence of legal standards, and/or lack of mechanisms for enforcing whatever legal standards that exist, mean that quality control is left almost exclusively to the bottlers. Combined with the poor quality control practices of the bottlers, it means that in many cases the consumers are not getting safe bottled water for drinking which they are paying. The general perception is that the quality and taste of bottled water is invariably better than tap water, which is not always true.

In many developing countries, quality standards do exist for bottled drinking water, but adherence to the standards is purely voluntary. Equally, since there are no specific regulatory requirements for establishing a bottled water plant, anyone who wishes to construct one can do so without indicating the source of water, technology used to purify it and the final quality of water bottled. Absence of regular, or any, monitoring by the competent authorities further gives the bottlers a free hand. Accordingly, the quality of bottled water is basically unknown in most developing countries.

Concluding Remarks

On the basis of the above analysis, it is evident that the provision of clean water, sanitation and disposal of stormwater to all the residents of the large urban centres of developing countries will be one of the major challenges of the 21st century, the magnitude and the complexity of which no earlier generation has had to face. Regular rhetoric at various international fora will not resolve this difficult and complex problem. The world at present has really two choices: to carry on as before with a 'business as usual' attitude which can only contribute to incremental changes that would endow the current and the future generations with a legacy of inadequate water supply and sewage services, or continue in earnest an accelerated effort to change radically the mind-sets of the decision-makers and the water managers so that the people in urban centres have access to safe drinking water and sanitation facilities within one generation. It would not be an easy task to accomplish, but given adequate political will and efficient management, it can be achieved. One is reminded of the warning of William Shakespeare that "men at some time are masters of their fates. The fault dear friends is not in our stars but in ourselves that we are underlings".

References

Asian Development Bank (2003) *Asian Water Supplies: Reaching the Poor* (Manila: Asian Development Bank).

Biswas, A. K., Jayatilaka, R. & Tortajada, C. (2005) Social perceptions of the impacts of Colombo water supply projects, *Ambio*, 34(8), December, pp. 639–644.

Rodriguez, R., O'Neal, G. (2006) Water quality management: North American Development Bank experience, in: A. K. Biswas, C. Tortajada, B. Braga & D. Rodriguez (Eds) *Water Quality Management in the Americas*, pp. 167–177 (Berlin: Springer).

United Nations Population Division (2002) *World Urbanization Prospects: The 2001 Revision*, Publication ESA/P/WP.173 (New York: United Nations Secretariat).

Megacities, Development and Water

OLLI VARIS

Introduction

We all learned at school that population growth would be the ultimate threat to the planet as well as the driver for its development. But while it is true that the world's population has grown, future projections are coming down relatively rapidly. In 1992, the United Nations Population Bureau predicted that the world's population in 2050 would be slightly more than 10 billion, but 10 years later the projection had dropped to 8.9 billion. This means that mankind has more time to deal with the problems that emerge from population growth: in the 1992 UN projection, the figure of 8.9 billion was expected to be reached before 2030, while the more recent estimate is that the population will not be at that level until 2050. This is a big change indeed.

Hence, the focus in the population discussion has gradually shifted towards urbanization. A landmark in this shift was the 1996 UNCHS (HABITAT) Conference on Human Settlements, which drew attention to the prediction that half of the world's 6 billion people would be living in urban areas, and half in rural, by the turn of the millennium. The rural population would remain almost stagnant after that, while urban population would grow from 3 to 5 billion between 2000 and 2025. In fact, almost all the urban population growth would take place in developing countries, meaning that their urban population would more than double over that period.

This implies that in 2025, at the global level, rural areas will have to supply food, energy and other commodities to towns and cities that will house almost twice as many people as the rural areas do. The global cycles of carbon, nutrients, water and other substances will be notably distorted, while markets perturb age-old social systems.

The aim of this paper is to outline the world's urban development, and in particular the growth of the largest agglomerations, known as megacities. This analysis draws attention to nine that have been selected to provide a comprehensive analysis of water management in megacities. The cities and the evolution of their populations are presented in Figure 1.

Urbanization, Megacities and Development: A Global Outlook

Let us continue to adopt the perspective of mankind only one generation ahead in time—the shortest possible span for any consideration of sustainable development. Each year, the world population grows by some 70 million. Almost all of this is urban growth in developing countries, although mainly it is due to migration; fertility rates are far lower in urban areas than in rural ones (UN, 2002). Urbanization and rural and urban population by continents is shown in Figure 2.

In Africa and in Asia, the proportion of urban population to rural is around 1:3, while in all the other continents it is more than 2:3. Therefore, urbanization is expected to be most rapid in Asia and in Africa (Figure 1). In China, it has been estimated that by 2025 the urban population will have grown by 350 million. China's cities currently house 'only'

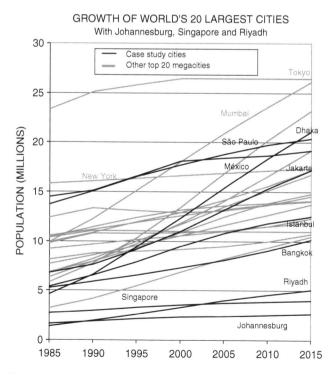

Figure 1. World's biggest cities in 2015 with Johannesburg. Population growth between 1985 and 2015 is shown. Names of the study cities are highlighted in bold type. *Source*: UN (2002).

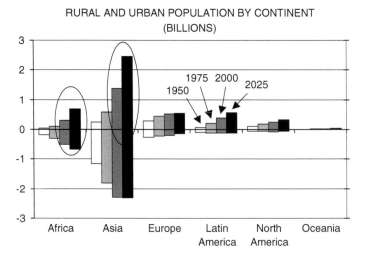

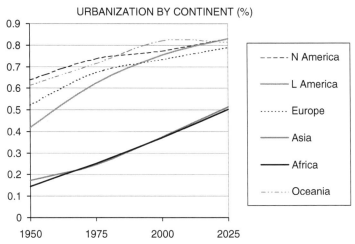

Figure 2. Rural and urban population by continent. *Note*: In the upper graph, the upper columns represent urban population and the lower ones rural population. *Source*: UN (2002).

500 million, and they already face severe problems of environmental and resource degradation.

Megacities

Urbanization is seen most dramatically as the growth of large urban agglomerations. The UN uses the term 'megacity' for one that has more than 5 million inhabitants; a 'big city' has more than 1 million. These terms are used also in this text. Whereas in 1985, there were 31 megacities on this planet, the number had grown to 40 by 2000, and is expected to grow to 58 by 2015.

The UN Population division regularly publishes statistics on the world's large urban agglomerations, including all those that had more than 750 000 inhabitants in 1990. The statistics, as of 1999, included 488 such cities. In what follows, some of the most important characteristics of these statistics will be compiled.

Urbanization in developing countries is notably faster than in the industrialized world, and many of the megacities grow at very rapid rates. Much of this urban growth occurs uncontrolled, with government controls having at best a limited impact. The fastest growing megacities are expected to increase more than four-fold within 25 years. These include Dhaka (Bangladesh), Lagos (Nigeria), Guatemala (Guatemala) and Jinxi (China). The dimension of the phenomenon is striking, whether it is considered from the standpoint of megacity growth, increased urbanization, or growth of the urban population.

Figure 3 provides a summary of population sizes and the evolution over time of the world's big cities. It also indicates the cities that have been chosen for comprehensive analysis in this study.

The world's total big-city population was slightly more than 700 million in 1985. By 2000, it had surpassed 1.1 billion, and by 2015 it is expected to approach 1.5 billion, double the level of 30 years earlier. In megacities, population growth has been equally dramatic and is expected to continue be so; from 270 million in 1985, it had grown to 400 million by 2000, and it is expected to reach 600 million by 2015 (Table 1).

Megacity growth is fastest in South Asia (Figure 3, Table 1). Whereas in 1985, the region had 19 million people in megacities, the number was 29 million in 2000, and it is expected to reach up to 65 million by 2015, three-fold growth in 30 years. In 2000 South Asia's megacities housed a mere 6% of the region's population, whereas in Latin America, the comparable figure was 15% and in East Asia, 24%. Southeast Asia is in a fairly similar situation to that of South Asia. However, the most important difference is that the megacities are smaller than those of South Asia, and are expected to grow more slowly.

China has the largest urban population of all these regions, currently some 200 million living in cities with more than one million inhabitants. However, its megacities are growing much less rapidly than those of South Asia. In fact, China's most extensive urbanization is taking place in provincial centres that typically have less than 5 million people. China has 108 cities with populations in the 1–5 million range, and most of them grow much faster than the nation's megacities. However, statistics never tell the full story. Some recent studies have revealed that the number of unregistered, mainly young, migrants in Chinese cities is very large. Both Shanghai and Beijing, for example, are estimated to have as many as 5 million inhabitants more than the official statistics indicate (Söderlund *et al.*, 2005).

Latin America has few, but very large, megacities. Their growth is also slowing down, and the biggest increase is to be expected in cities of less than 5 million inhabitants.

Europe will have a much slower growth rate of cities than most other regions. Its big city population of some 170 million is now almost equal in size to that of Latin America and the Caribbean. Europe's only rapidly growing megacity is Istanbul, which is just passing the milestone of 10 million inhabitants. Other megacities have very modest growth rates. North America's megacities grow at a very slow pace. Oceania has no megacities, and only a handful of big cities. East Asia has huge megacities, and their share of total population is higher than anywhere else, almost one-quarter (Figure 4). However, their growth is very slow, as is the whole urbanization process in East Asia.

In turn, cities of the Middle East have grown very rapidly in recent decades, and continue to do so. Megacities had only 4% of the total population in 2000, and their growth continues to be considerable.

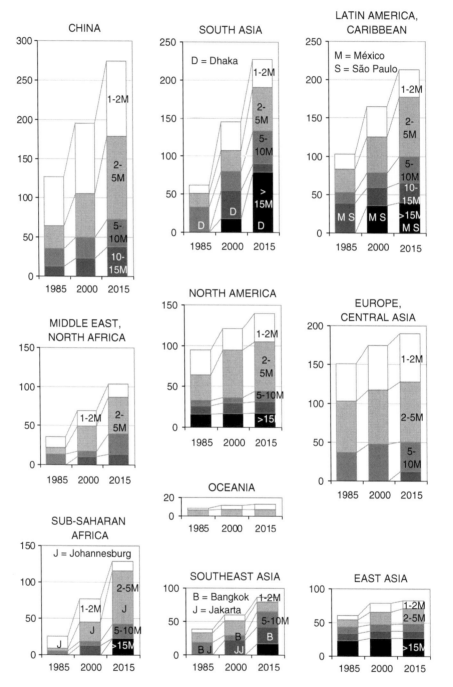

Figure 3. Population of world's cities by region. The population is classified by the size of the cities (1–2 million, 2–5 million, 5–10 million, 10–15 million and cities with more than 15 million inhabitants. *Source*: UN (2002).

Table 1. Megacities of the world: population in 1985, 2000 and 2015 (millions). These cities are the ones that are predicted to have over 5 million people in 2015 by UN (2002)

	Country	City	1985	2000	2015	Growth in 30 years (times)	Income class
China	China	Shanghai	12.4	12.9	14.6	1.2	Low
	China	Beijing	9.8	10.8	12.3	1.3	Low
	China	Tianjin	8.1	9.2	10.7	1.3	Low
	China	Chongqing	2.8	5.3	8.9	3.2	Low
	China, HK SAR	Hong Kong	5.1	6.9	7.7	1.5	High
	China	Wuhan	3.5	5.2	7.4	2.1	Low
	China	Shenyang	4.2	4.8	5.7	1.4	Low
	China	Jinxi	0.9	2.8	5.4	6.0	Low
	Total		*46.8*	*57.9*	*72.6*	*2.1*	
East Asia	Japan	Tokyo	23.3	26.4	26.4	1.1	High
	Japan	Osaka	10.4	11.0	11.0	1.1	High
	Republic of Korea	Seoul	9.5	9.9	9.9	1.0	High-middle
	Total		*43.2*	*47.3*	*47.4*	*1.1*	
Europe	Turkey	Istanbul	5.4	9.5	13.0	2.3	Low-middle
	France	Paris	9.1	9.6	9.7	1.1	High
	Russian Federation	Moscow	8.6	9.3	9.4	1.1	Low-middle
	United Kingdom	London	7.7	7.6	7.6	1.0	High
	Germany	Essen	6.2	6.5	6.6	1.1	High
	Russian Federation	Saint Petersburg	4.8	5.1	5.1	1.1	Low-middle
	Total		*41.8*	*47.7*	*50.9*	*1.4*	
Latin America	Brazil	São Paulo	13.8	17.8	20.4	1.5	High-middle
	Mexico	México	14.5	18.1	19.2	1.3	High-middle
	Argentina	Buenos Aires	10.5	12.6	14.1	1.3	High-middle
	Brazil	Rio de Janeiro	9.2	10.6	11.9	1.3	High-middle
	Peru	Lima	5.1	7.4	9.4	1.8	Low-middle
	Colombia	Bogotá	4.4	6.3	8.0	1.8	Low-middle
	Chile	Santiago	4.1	5.5	6.6	1.6	High-middle

Table 1. Continued

Country	City	1985	2000	2015	Growth in 30 years (times)	Income class
Guatemala	Guatemala City	1.1	3.2	5.3	4.8	Low-middle
Brazil	Belo Horizonte	2.9	4.2	5.0	1.7	High-middle
Total		*65.6*	*85.7*	*99.9*	*1.9*	
Middle East						
Egypt	Cairo	7.7	10.6	13.8	1.8	Low-middle
Iran (Islamic Republic of)	Teheran	5.8	7.2	8.7	1.5	Low-middle
Iraq	Baghdad	3.7	4.8	6.8	1.9	Low-middle
Egypt	Alexandria	2.8	4.1	5.5	1.9	Low-middle
Saudi Arabia	Riyadh	1.4	3.3	5.1	3.6	High-middle
Total		*21.4*	*30.0*	*39.9*	*2.9*	
North America						
United States	New York	15.8	16.6	17.4	1.1	High
United States	Los Angeles	10.4	13.1	14.1	1.3	High
United States	Chicago	6.8	7.0	7.4	1.1	High
Canada	Toronto	3.4	4.7	5.3	1.6	High
Total		*36.4*	*41.4*	*44.2*	*1.3*	
South Asia						
India	Mumbai	9.9	18.1	26.1	2.6	Low
Bangladesh	Dhaka	4.7	12.3	21.1	4.5	Low
Pakistan	Karachi	6.3	11.8	19.2	3.0	Low
India	Calcutta	9.9	12.9	17.3	1.7	Low
India	Delhi	6.8	11.7	16.8	2.5	Low
India	Hyderabad	3.2	6.8	10.5	3.3	Low
Pakistan	Lahore	3.5	6.0	10.0	2.9	Low
India	Chennai	4.7	6.6	9.1	1.9	Low
India	Bangalore	3.4	5.6	8.0	2.4	Low
India	Ahmedabad	2.9	4.2	5.8	2.0	Low
Afghanistan	Kabul	1.2	2.6	5.3	4.4	Low
India	Pune (Poona)	2	3.5	5.1	2.6	Low
Total		*58.5*	*102.1*	*154.3*	*4.7*	

Southeast Asia	Indonesia	*Jakarta*	6.8	11.0	17.2	2.5	Low
	Philippines	Metro Manila	6.9	10.9	14.8	2.2	Low-middle
	Thailand	Bangkok	5.3	7.3	10.1	1.9	Low-middle
	Viet Nam	Ho Chi Minh City	3.7	4.6	6.2	1.7	Low
	Myanmar	Yangon	2.8	4.2	6.0	2.2	Low
	Indonesia	Bandung	2.1	3.4	5.2	2.5	Low
	Viet Nam	Hanoi	2.9	3.7	5.1	1.8	Low
	Total		*30.4*	*45.1*	*64.8*	*3.4*	
Sub-Saharan Africa	Nigeria	Lagos	5.8	13.4	23.2	4.0	Low
	Dem. Rep. Of the Congo	Kinshasa	2.8	5.1	9.4	3.4	Low
	Ethiopia	Addis Ababa	1.5	2.6	5.1	3.4	Low
	Côte d'Ivoire	Abidjan	1.7	3.3	5.1	3.0	Low
	Total		*11.8*	*24.4*	*42.7*	*7.2*	
World	*Total*		*356.0*	*482.0*	*616.7*	*2.3*	

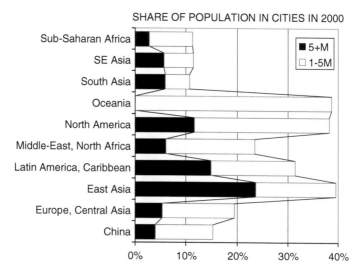

Figure 4. Percentage of population living in big cities (1–5 million) and megacities (5+ million) by region. *Source*: UN (2002).

Sub-Saharan Africa is undergoing a massive urbanization process, which is obviously no more than a prelude to what will follow in the coming decades. The region's level of urbanization is still very low (one-third, Figure 2). In 1985, cities of more than 1 million people had 26 million inhabitants, whereas the prospective number for 2015 is 129, a five-fold increase in 30 years. This level of growth is far more dramatic than in any other part of the world. However, there are still only two megacities in Sub-Saharan Africa. Lagos in Nigeria is by far the larger, while Kinshasa in the Democratic Republic of the Congo has just passed 5 million.

Urbanization and Development

It seems that the most fundamental impact of urbanization will be on low-income countries, where the pace of change will be very fast and the duration long. In terms of population share, India and China are in pole position; they have roughly two-thirds of the world's low-income category population. However, China is about to jump to the low-middle income category, thereby halving the population of low-income countries.

Not all low-income megacities grow rapidly. In particular, the biggest Chinese Cities, Shanghai, Beijing and Tianjin, have relatively low growth at present. However, forecasts for China may turn out to be far too conservative. In any event, a majority of low-income megacities will double or triple their populations between 1985 and 2015, posing enormous challenges in many ways, particularly for Lagos and Dhaka, which are already very large, have deficient infrastructure and services, undeveloped economies, and continue to grow much faster than other megacities with more than 12 million inhabitants (Figure 5).

Most of the low-middle income megacities are expected to grow 1.5 to two-fold in the period 1985–2015. Guatemala City, driven by its history of civil conflict, is a clear front-runner because of its enormous growth rate.

It is interesting to look at development indicators for different parts of the world and relate them to urbanization. First, there will be a look at the economy. The urbanization level of the

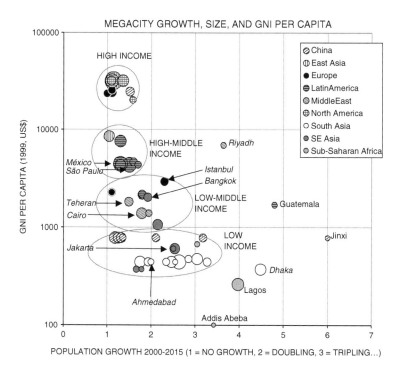

Figure 5. World's megacities: population growth rate against Gross National Income per capita. Bubble size is relative to population size of each city. *Sources*: UN (2002) and World Bank (2001).

world's macro-regions seems to have a close connection to their Gross National Income (Figure 6a). The World Bank (2003) statistics for 1980–2000 reveal that most regions have been able to improve their economies roughly at a pace with urbanization. However, there are some exceptions. In Africa and the Middle East, economies have declined whereas urbanization has been rapid. In Latin America and the Caribbean, economic development has been unable to keep pace with urbanization. In these regions, the social and economic problems that result from urbanization can be expected to be more severe than elsewhere in the world.

The pace of economic development and level of education are important factors. Southeast Asia and China have strong economic development, and illiteracy rates at the same level as those in Latin America and Caribbean (Figure 6b). However, while Southeast Asia and China had only 40% of the population in urban areas by 2000, some 80% of Latin Americans were living in cities by then. In the Middle East and North Africa, literacy levels are far lower than in either of those regions, but urbanization is relatively high, at more than 60%.

Figures 6b and 6c indicate the two different paths to urbanization, one in which people learn to write before massive urbanization takes place, and the opposite. Obviously, regions that take the former path will have a smoother way ahead in coping with the challenges of urbanization. Economic development is strongly dependent on levels of education (Figure 6d), and there is no way to prosperity if people are denied learning. It will be interesting to see which of the two ways South Asia and Sub-Saharan Africa take in the coming years and decades. Maybe they will find a middle way, combining aspects of both. For a more detailed statistical analysis of Southeast Asian, West African and Latin American countries, see Haapala (2002).

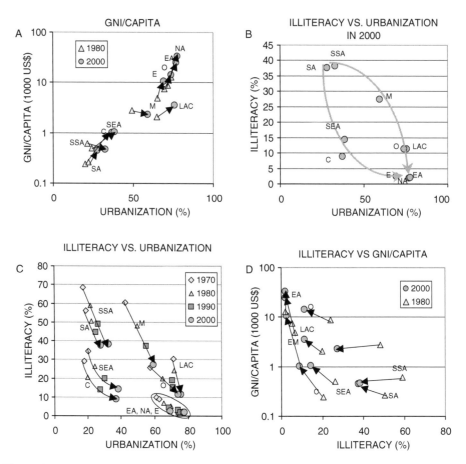

Figure 6. Urbanization, economy and illiteracy: developments by region in the period 1980–2000. *Source*: World Bank (2003). NA = North America, EA = East Asia, O = Oceania, E = Europe, LAC = Latin America and Caribbean, M = Middle East and North Africa, SEA = Southeast Asia, C = China, SSA = Sub-Saharan Africa, SA = South Asia.

An even more homogeneous group of megacities is those, in Latin America plus Seoul in Korea, which are located in high-middle income countries. With a projected growth rate of 15–20% between 2000 and 2015, they are no longer growing much faster than the world's 10 high-income cities (Figure 5). These statistics are, of course, indicative only. A serious shortcoming is that the UN agencies and the World Bank publish primarily country-level statistics, such as the GNI data used in this analysis. However, the cities are often very far from national averages in many ways, including income indicators. It is also questionable whether some of the growth rates, such as those of Jinxi, are realistic at all.

From National to a Global Context

In many countries the biggest urban centres grow faster than other cities. In Thailand, the growth rate and wealth of Bangkok greatly outpace those of other Thai cities (Figure 7).

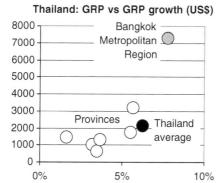

Figure 7. Thailand's and Bangkok's economy. Bangkok's strong economy in comparison to the economy of the rest of Thailand. Annual rates in the period 1989–93 are shown. GRP = Gross Regional Product, in US$. *Source*: Pednekar (1997).

Manila contains 79% of the Philippines' manufacturing employment. A counterexample is China where hundreds of middle-sized cities grow now much faster than the biggest cities.

As a result, many capitals have been placed under direct government administration, almost to the point of being quasi-independent (e.g. Jakarta, Bangkok Metropolitan Area, Metro Manila, Federal Territory of Kuala Lumpur, México DF, Beijing, Tianjin, Chongqing, etc.). These cities evidently want to be among those that rise from being peripheral to semi-peripheral, or even core, socio-economic units on a global scale (cf. King, 1991). Most developing regions have a pressing need for such centres. Megacities that are unable to become core socio-economic units on a global level tend to have major problems in meeting their many needs. Among the world's megacities, Lagos is perhaps the most dramatic example.

Cities, above all the biggest ones, are the interface between a country or region and the globalizing economy and culture. With their concentrations of capital, educated inhabitants and other resources, they may be networked more to each other than to their hinterlands. Owing to their economic links to the world market, they often prefer to purchase food and other commodities there if the market structure or prices cannot compete in the country or region in which they are located. This is obviously one outcome of the frequently made claim that nation–states are to a certain extent ceding influence in today's world (Hewett and Montgomery, 2001).

Pushed and Pulled Migrants

It is one thing if a migrant with a basic education and access to a formal-sector job and the possibility of higher education is "pulled" to a city, but quite another if he or she is "pushed" out of rural areas with no education or other means of making a decent living in urban society. The rules of the game are very different because traditional skills are of little use in a big city.

The division of society into formal and informal sectors is most evident in the rapidly urbanizing centres of developing countries. Besides presenting a major challenge to any aspect of water and infrastructure development, it equally involves issues of safety, revenue collection, health issues and skilled labour.

The hope tends to be that the growth of industry and urban services will somehow absorb the excess immigrant labour force. However, this usually remains a pipedream for all but a handful of people. Even in the most successful countries in this respect, such as Brazil and Thailand, only one-fifth of the population is engaged in industry. The informal sector has to absorb 20 to 70% of the urban workforce (Todaro, 1997). According to Langman (2003) 8 out of 10 new jobs in Latin America are in the informal economy. In the Gambia, the informal sector employs 30% of the rural workforce, and 60% in the cities (Esim, 1996). The Gambia's urban population grows by 6.8% a year, and most of it moves into the informal sector.

Urban poverty is massive and increasing. Whereas the number of urban poor in the world today is reckoned to be about 1 billion, it is expected to double by 2030. After the 2004 HABITAT conference, the humble UN target defined in the Millennium Development Goals of reducing the number of slum dwellers by half is likely to remain a dream. What seems more likely is that the number will double. Slum upgrading policies, particularly in bigger urban agglomerations, are therefore expected to become very important in the coming decades, with major implications for the water sector.

In fact, UN-HABITAT (2003a) defines a slum dweller in a very interesting way, as a combination of the lack of the following: improved water supply, improved sanitation, sufficient living area, durable housing and secure tenure. The provision of water and sanitation services is, therefore, one of the keys to upgrading slums and reducing poverty.

Megacities and Water

Location with Respect to Water

The megacity population is surprisingly evenly distributed across different climatic zones (Figure 8). A particular challenge is that roughly one-third of the megacity population lives in arid areas, and 10% inhabit desert megacities (Cairo, Lima, Baghdad, Riyadh, Alexandria, Karachi).

Some 60% of megacity inhabitants live in coastal areas, one-third of them on the deltas of major rivers. The share of inland cities is remarkable. In 2025, almost quarter of a billion people are expected to have their home in inland-megacities. These cities face particular challenges in meeting water needs (Abderrahman, 2000).

According to the IFRC (2002), the number of people exposed to floods tripled between the 1970s and 1990s, and stands at about 2 billion today. This is chiefly due to the concentration of people in the floodplains of big rivers or cyclone-prone coastal areas, particularly in Asia. Wuhan, Dhaka and Bangkok are examples of rapidly growing cities that are extremely flood-prone. The informal settlements of cities are particularly vulnerable.

Waterways are a major transport facility in most megacities, except for the high-altitude ones and some others that lie inland.

All water-management theories and agendas say that measures should be taken in a basin-wide context. But these theories barely take account of megacities. Megacities need water, food and energy, and 63% of their inhabitants are not located in a major river basin that would afford them the lowest water scarcity limit of $1000 \, m^3$ per capita per year. In 2015, the population living in these circumstances will be 387 million.

The bulk of the world's economic activity will be concentrated in megacities, where water management does not easily fit into the basin-wide context. They operate in an

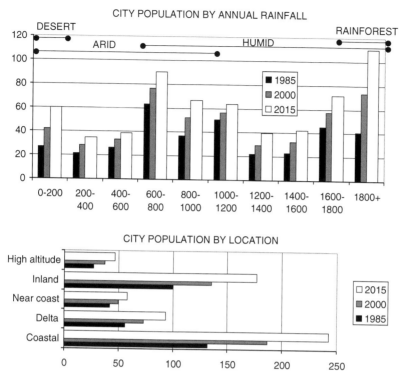

Figure 8. Development of megacity population by available annual rainfall and by location with respect to the sea. *Source*: Data from various sources.

environment that is far more global and driven by trade as well as by politics than the somewhat simplistic basin-wide argument would suppose.

The importance of groundwater aquifers is crucial to many megacities (Table 2). Almost one-half of the megacity population may have access to highly productive aquifers.

Water supply and sanitation. Estimates of the availability of adequate water and sanitation services in the world's urban areas vary enormously. Table 3 shows two different estimates (cf. UN-HABITAT, 2003b). One of them claims that in Africa, some 15–16% of urban dwellers lack the provision of improved water and sanitation. The other one (WHO and UNICEF, 2000) says that the numbers are 35–50 and 50–60%, respectively. For Asia and Latin America, the numbers have an equally wide discrepancy. The comparison of these two statistics only shows the inaccuracy of the data with which we have to work.

It is clear that the problem is vast, but its true dimension seems to be highly unclear. A reduction in the number of people with inadequate water supply is one of the UN Millennium targets, but statistics seem to disagree even on the size of the problem.

According to Hewett & Montgomery (2001), the world's megacities tend to have somewhat better coverage of water and sanitation services than other urban areas. Smaller cities tend to be less well served than larger ones throughout the developing world.

Table 2. World's megacities with respect to annual rainfall and availability of groundwater aquifers (rainfall from various sources, aquifers after UNESCO, 2003)

Annual rainfall (mm)	Major groundwater basin with highly productive aquifers	Complex structure including some important aquifers	Local and shallow aquifers	Saline groundwater
0–200	Riyadh, Baghdad	Teheran, Los Angeles	Lima	Cairo, Alexandria, Karachi
200–400	Santiago		Kabul	
400–600	Lahore, Moscow, México			
600–800	Tianjin, Beijing, London, Essen, St Petersburg, Paris, Delhi, Shenyang	Pune	Toronto, Jinxi	
800–1000	Ahmedabad, Buenos Aires	Istanbul, Hyderabad, Bangalore, Bogotá	Chicago	
1000–1200	Shanghai		Osaka, New York, Rio de Janeiro, Chongqing	
1200–1400	Seoul, Chennai, Bangkok, Belo Horizonte, Kinshasa, Wuhan	Guatemala	Addis Ababa	
1400–1600			São Paulo	
1600–1800	Hanoi, Tokyo	Jakarta, Bandung	Abidjan, Hong Kong	Calcutta
1800 +	Lagos, Yangon	Metro Manila, Mumbai		Ho Chi Minh, Dhaka
Average rainfall	975	1147	1047	1086
Population in 2000 (millions)	228	99	99	56
Population in 2015 (millions)	278	136	120	83
Growth 2000–2015	22%	38%	22%	48%

Note: Study cities in italics.

Table 3. Number and percentage of people without appropriate/improved water and sanitation services in Africa, Asia, and Latin America. Two different estimates

	Assessment of WHO and UNICEF (2000)		Assessment of UN-HABITAT (2003b)	
	Water millions (%)	Sanitation millions (%)	Water millions (%)	Sanitation millions (%)
Africa	44 (15%)	46 (16%)	100–150 (35–50%)	150–180 (50–60%)
Asia	98 (7%)	297 (22%)	500–700 (35–50%)	600–800 (45–60%)
Latin America and Caribbean	29 (7%)	51 (13%)	80–120 (20–30%)	100–150 (25–40%)

Sources: WHO and UNICEF (2000); UN-HABITAT (2003b)

Some comparative data is presented in Figure 9 for illustrative purposes. The baseline is that the priority seems to be the water supply, and thereafter comes sanitation and waste management with a considerable time lag. The service level seems to have a certain relation to the income level of different regions and cities, although many exceptions exist. Among them, the fairly high service level of China can be mentioned, although China is considered as a low-income country. Again, the statistics may well be biased in this respect.

Indeed, the special concern is due to the very low service level of Africa. Although the statistics in Figure 9 include only access to piped water systems and sewerage connections, any other indicator would tell the same story: the infrastructure of African cities is much less developed than those in other parts of the world. In fact, in North Africa, the situation is much better. Were the statistics to include only Sub-Saharan Africa, they would make much more sober reading.

The dimension of the task of infrastructure provision in rapidly growing urban areas can be illustrated with an example from China, whose urban areas produced 35 km³ of waste-water in 1997. This is expected to grow to 650 km³ by 2010, and to 960 km³ by 2030. The treatment level was 11% in 1997, and the target is to achieve 40% by 2010 (Oyang & Wang, 2000). The difficulty in meeting this target becomes clear inform a comparison with Germany (van Riesen, 1999). In 1996, Germany treated 22% more waste-water (in cubic metres) than China. In order to reach the target of 40% by 2010, China will have to build one-third of German's present capacity each year.

Every individual needs a certain supply of water each day. Therefore, water is a potentially lucrative commodity with which to do business, particularly among urban squatters, where traditions and old customary rights to water have faded away. The term 'hidden water economy' is often used in this context.

In Karachi, Pakistan, the poor must pay up to 40 times the official price for water from water vendors. The middle and upper sections of the population (the formal sector dwellers) receive water piped into their homes at the official rate (Baloch, 1999). Cairncross (1990) estimates that 20 to 30% of all urban inhabitants of the Third World are dependent on water vendors. The prices amount to 100 times the official tariff (Bhatia & Falkenmark, 1992). People's response to this dilemma can be disastrous. In Karachi, vendors sell water (with no quality control) at highly inflated prices. Private wells are common, and water drawn from them originates from diverse urban sources, partly from leaking sewerage pipes. Health risks are high (Rahman *et al.*, 1997). In addition, illegal tapping of water from leaking municipal water pipelines, even by damaging pipes, is common. However, it must be realized that the marginal cost of

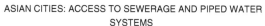

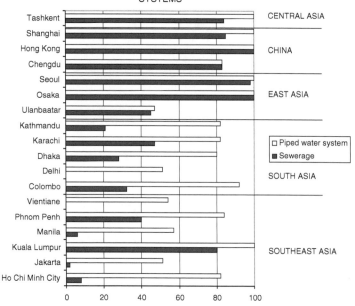

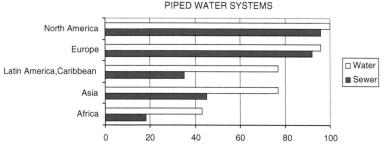

Figure 9. Some indicative data on water infrastructure. *Sources*: ADB (2003) (top) and UN-HABITAT (2003b) (bottom).

providing water to areas without an adequate infrastructure is often far higher than the price paid by people with appropriate services and the cost to society of providing services to slum areas would not perhaps be much lower than the tariffs of the water vendors (Lundqvist *et al.*, 2003).

Food

Food production is the biggest consumer of water in the world, accounting for 70% of the total, and more than 90% in many developing countries. Since almost all population growth already takes place in the cities and will continue to do so in coming decades, agriculture faces a huge challenge in achieving greater efficiency and becoming more market-driven.

Equally, the escalating food markets will mean swelling the 'virtual' water trade. The concept of virtual water has entered the discussion of water and food trade in the sense of water that is needed to produce food, which is then traded. If for instance two tons of water is required to produce one kilogram of rice, then the virtual water content of each traded kilogram of rice is the equivalent of two tons.

The following simple calculation helps in comprehending the order of magnitude of virtual water imports to cities. Let us assume that a city dweller in a high or high-middle income country has a diet that consumes on average 5500 litres of water per day, and those in lower-middle or low-income countries 4500 and 4000 litres, respectively. Multiplying these figures by the population of megacities in the given income classes and across world's regions, we obtain the numbers shown in Figure 10. Comparing these figures with the Nile's annual flow at Aswan, which is 55 km^3, it can be seen that the 10 megacities of Latin America import roughly three times the flow of the Nile in terms of virtual water.

Hoekstra & Hung (2002) estimated that all the world's international trade would add up to 1000 km^3. Interestingly enough, if the virtual food imports to megacities are added up, the result is the same figure.

The world and its cities are thirsty for water, but too often it is forgotten that they are far more hungry for water. Table 4 shows that, whereas all of mankind drinks some 4.4 km^3 per year, it eats some 10 000 km^3 of water in the same time. The population growth of 70 million a year eats roughly twice the flow of the Nile.

Figure 11 shows the comparison of the size of megacity population and that of the population to which food was or will be imported in the years 2000 and 2015, respectively. The food (cereal) import calculation was made on the basis of the IMPACT model results by Rosegrant *et al.* (2001, 2002).

Roughly speaking, China imports cereal to almost as many people as dwell in its megacities. This does not necessarily mean that China feeds its megacities with imported food, but it tells us that China's food deficit is equal to the size of its megacity consumption. In South Asia, the food deficit is expected to grow very close to megacity size by 2015. In 2000 it was about half of that.

In South Asia, megacity food consumption equalled the food deficit in 2000, but is expected to double by 2015 whereas food imports are not expected to grow.

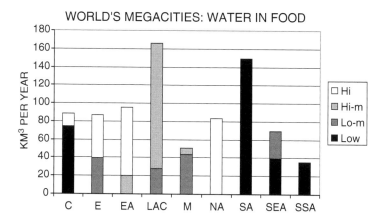

Figure 10. Virtual water import to megacities in the year 2000

Table 4. Indicative comparison of water used to produce nutrition to people and water used for drinking purposes

	Drink	Use in the household		Food		Unit
		Rural areas	Urban areas	Green & blue water	Blue water	
1 person	3	50	250	4500	1400	Litres per day
	1	20	90	1650	500	m^3 per year
6 billion people	4.4			10 000	3000	km^3 per year
	0.04%			84%	26%	Of stable flow
	0.01%			24%	7.5%	Of continental flow
	140			310 000	97 000	m^3 per second
70 million urban each year	0.07		6.3	115	35	km^3 per year
	0.001		0.1	2	0.6	times the Nile

In the Middle East and most of Sub-Saharan Africa, domestic food production is far too low to feed the large urban areas. Sub-Saharan Africa imports cereal to almost three times the size of the population that lives in its megacities. Both imports and megacity population are expected to double between 2000 and 2015. In the case of the Middle East and North Africa, the imports are even larger in relation to megacity population. These regions feed the bulk of their urban population with imported food: in 2000 they imported cereal for 126 million people whereas their total urban population was 175 million.

The big trade-related question is whether the countries will feed their cities themselves or whether they import food. In either case, a huge expansion of market-driven agriculture will be needed. But which regions are able to feed their megacities with imported food? Most imports come from high-income countries, as well as from Latin America and Southeast Asia. Presumably, the Middle East can afford its food imports with relative ease, but what

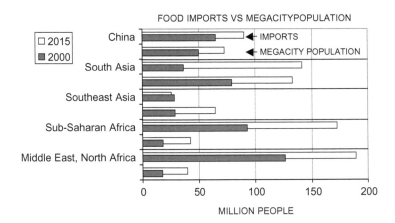

Figure 11. Comparison of megacity population and the cereal net imports calculated to correspond the number of people to which the food is imported. The major food-importing regions of the world. *Sources*: Population data UN (2002); food import calculations are based on IFPRI's IMPACT model; Rosegrant *et al.* (2001, 2002).

about Sub-Saharan Africa? With low level of export earnings, the region's food imports contribute to the strongly negative trade balance, which leads to ever-growing debt.

Energy

Megacities are major consumers of energy. Lagos accounts for 45% of all Nigeria's energy consumption, although it has 11% of the nation's population. Energy consumption per capita in Lagos is thus seven times greater than in the rest of the country. In China, urban energy consumption is about three times greater than in rural areas, and growing much faster. Bangkok accounts for one-third of Thailand's electricity consumption.

Two major sources of energy, biomass and hydropower, are very closely tied with water. In fact, they are the most important renewable sources of energy. Biomass production consumes large quantities of water. World bioenergy production consumes around $1300 \, km^3$ water per year. This is about one-eighth of the water used for food production and close to the amounts of domestic water consumption or industrial water consumption at a global level (Varis, 2005a).

Biomass is particularly important in developing countries. In South Asia, biomass accounted for 40% and hydropower 4% of all energy production in 2000, whereas in North America, their shares were 5% and 4% respectively.

For megacities, hydropower would be a preferable source of energy, since it creates very few environmental problems within the cities. The effects are born by others—that is the controversy—but the environmental impact of hydropower is often relatively small in comparison to many other ways of producing energy. In the case of biomass burning, the local effects are typically extremely problematic in congested urban areas. Dhaka already has the poorest air quality of the world's megacities (ADB, 2004). Its energy infrastructure is underdeveloped, and air pollution is due in part to the open burning of biomass.

Megacity growth thus presents a challenge for the energy sector. Traditional renewable sources of biomass energy are ill-suited to urban areas. As the cities grow, the need for energy is typically covered by fossil sources such as coal, oil and gas. The share of hydropower remains typically within a range of 5–15%. Even in hydropower-rich Latin America, dams produce only 13% of all energy (Nakicenovic *et al.*, 1998). Urbanization increases pressure for the development of hydropower and commercial, industrially produced biofuels. However, the latter, are still waiting for a major breakthrough, and their share of global production is expected to remain below 5% for several decades to come.

The urban water sector also consumes huge amounts of energy. Tortajada (2003) estimates that in Mexico City Metropolitan Area, some 20% of electricity consumption is spent on the pumping of water and wastewater.

Environmental Degradation

Whereas in traditional rural societies, organic matter, nutrients and other material are largely recycled, this is not the case in urban agglomerations. Massive amounts of goods and materials are transported over long distances to reach urban users. However, they are not returned to the originating ecosystems afterwards, but discharged in the immediate environment of the large-density populations. There they create problems.

Water is the major transporter of this waste and, as it becomes contaminated, its usefulness to humans and nature is hampered, often drastically. Air quality problems in most developing-country megacities are at the level that produces an immediate health risk. Surface and groundwater deterioration is detrimental to health.

The case of the extremely water-affluent megacity of Dhaka is revealing. Serious surface and groundwater pollution with detrimental effects on public health are a consequence of the massive shortcomings of infrastructure for water supply and sanitation, made worse by occasional and, often dramatic, flooding. Stormwater management systems have been developed but have failed to keep pace with the growing population, particularly in the eastern part of Dhaka, which has a population of three million (Ahmad & Kamal 2004). Several decades ago, the city was covered by a network of 24 canals and included a large area of natural wetland. This system was able to keep flood damage at a fairly low level, but the unplanned and largely illegal sprawl of the city ever since has led to a situation in which no proper stormwater infrastructure exists. The most important flood protection system today is the Dhaka Western Embankment which is able to keep about half of the city area virtually flood-free (Haq, 2006, this volume).

Groundwater has been over-exploited massively in the Bangkok area of Thailand, and land subsidence and saltwater intrusion have followed. Water distribution networks have been extended and pumping restricted since 1988, and some improvement has been noted. However, the rapid and continuing extension of the city and the consequent spread of groundwater use have nullified this positive development. Several recommendations have been presented, principally calling for institutional development, including legislation (Das Gupta, 2005).

Mexico City grew on the site of the ancient Aztec capital in a valley enclosed by mountains. This location was perhaps very favourable for an ancient town but is quite the contrary for a modern megacity. Rapid growth has caused a decline in the quality of life in the city, which has become overcrowded, immensely polluted and has serious shortages of such basic amenities as water. A mere 9% of Mexico City Metropolitan Area's wastewater is being treated. Tortajada (2003, 2006a, this volume) points out that despite the immense infrastructure, no strategies exist for the integrated management of water resources. As a result, severe shortcomings persist in maintenance and the systematic development of infrastructure. Occasional heavy rains have caused some rivers, now open sewers, to flood inhabited areas and people (usually poor) are exposed to health risks. Equally, cracks in sewer canals have caused wastewater to flood dwelling areas. The baseline is that construction of the infrastructure that is needed to bring more and more water to the metropolitan area is neither socio-economically feasible nor environmentally sustainable. Costs are skyrocketing and the benefits go largely to the well-off part of the population. Environmental deterioration in terms of lowering groundwater level, land subsidence and degradation, as well as the deterioration of surface and groundwater quality, works against sustainability of this megacity.

Johannesburg is located on a ridge at the head of two major international rivers, the Orange and the Limpopo. The city does not, therefore lie next to a major river that it could use, but instead its activities have international hydro-political consequences. The reason why the city has such a very unfavourable geographical location is due to its growth having originated around gold mines (Turton et al., 2006, this volume). The mines have caused environmental problems, particularly of groundwater quality, at a time when surface water problems due to municipal wastewater are escalating. In the South African part of the Limpopo Basin, for example, more than 1000 abandoned mines are a source of heavy metals, sulphur and associated problems of acidity that pose grave problems to local and international water users and to the natural environment. Municipal wastewater is the cause of growing eutrophication, which also has wide-reaching effects even on downstream countries, particularly Botswana.

Connection to Rural Development

Because most of urban population growth is due to migration from rural areas, one efficient way of reducing urban problems, including those involving water, is to promote rural development. China is an interesting case in this respect. For several decades, the government restricted urbanization, but recently the policy has been relaxing. At the same time two parallel processes are taking place.

On the one hand, the urban areas are witnessing unforeseeable economic growth that allows massive improvements in livelihoods and infrastructure; very large numbers of people have emerged from poverty. On the other hand, similar progress has not taken place in rural areas, and subsidies, which used to be very high, have been reduced. Rural areas are short of financial institutions services, and other market infrastructure.

As a result, the expanding coastal megacities have begun to import food from the world market and the new urban wealth fails to trickle down to the rural economy. Urbanization continues apace, and informal settlements have become considerable in size in Chinese cities. The sustainability of agro-ecosystems and rural livelihoods are also in question. Some recent studies reveal that the number of unregistered, mainly young migrants in Chinese cities is soaring, which results in a rapidly growing informal sector (Söderlund *et al.*, 2005).

Discussion and Conclusions

This paper has scrutinised the interface and interplay of two profoundly important aspects of mankind's global challenges of sustainable development. One is urbanization and the other is water. Urbanization, including the growth of large urban agglomerations, is the big demographic issue of the current and future decades. Virtually all population growth, now around 70 million each year, ends up in urban areas. Most of the growth is caused by the migration of poor rural people to urban areas, particularly in Asia and Africa. The less affluent a country, the more water is a determining factor in the economy, and the less society's capacity to cope with problems caused to the environment, including those related to water (e.g. Haq, 2006, this volume). Ironically, the poorest countries are those that are urbanizing at the fastest pace.

Particularly striking is the situation in the large urban agglomerations. The population of the world's megacities is growing by 9 million a year and now amounts to about half a billion inhabitants. These cities attract a rapidly increasing share of the world's economic activity, making them major hubs of the globalizing economy. They also challenge many of the prevailing paradigms of water management, since they interact in many ways across both jurisdictional and hydrological borders, while portraying extreme contrasts between the expanding slum population reliant on the informal economy and informal institutions, and the modern, rapidly globalizing sector.

Despite the importance of the informal sector, few assessments of water issues address them adequately. Yet the informal sector is the soaring and almost ignored urban water problem (Varis, 2001). Informal settlements and economic systems are the most vulnerable to economic cycles, and natural disasters such as floods and droughts. UN-

HABITAT (2004) promoted the idea of 'inclusive' cities that would be less markedly divided into social layers. Whereas this is a good idea, it faces a great challenge, since it seems that the world is not going towards mutual social responsibility but somewhere else. A more pronounced tendency at present is that modern sector services to the middle and upper-class dwellers may be strikingly similar whether they live in a town house in Beijing, Dhaka, Johannesburg, São Paulo or somewhere else. Yet, the sizes of the upper and middle class differ widely, being very small in cities such as Dhaka.

Macroeconomic growth is a prerequisite of a reduction in urban poverty. In a comparison, UN-HABITAT (2003a) maintains that urban poverty is far more sensitive to macroeconomic conditions than rural poverty. For instance, as the result of an economic recession in Latin America in the 1980s, per capita income decreased by 10%, and while the World Bank estimates that overall poverty rose by 17%, urban poverty escalated by 73% (ECLAC, 1999). Therefore, economic development is important in reducing urban poverty. Bangkok (Heinonen, 2004) and Chinese megacities such as Shanghai are good examples of the possibilities that rapid economic growth opens for environmental and infrastructure policies, as well as poverty reduction. Priority in the establishment of national and international policy is the prerequisite of such development. Johannesburg (Turton *et al.*, 2006, this volume) and Istanbul (Altinbilek, 2006, this volume) provide further examples. Improper infrastructure and environmental management, as well as social systems with too great a poverty gap, hamper economic development and care for the environment. Strong strategic planning is neglected too often. However, whereas the reduction of urban poverty is very difficult without economic development, economic growth does not necessarily lead to poverty reduction. Sound, inclusive socially-targeted policies are needed, many of them, such as improvements in education, being efficient only over a rather long-term perspective. UN-HABITAT (2003a) maintain that success in poverty reduction is relatively closely related to human development.

It is astonishing how often urbanization is disregarded in development studies and policy documents. For example, most UN and World Bank statistics draw no distinction between the urban and rural areas of countries, even though each is typically extremely different from the other. Equally, targets such as the Millennium Development Goals that largely rely on such statistics and benchmarks, are not sufficiently sensitive to urbanization.

Equally astonishing is that aspects other than supply and sanitation are very seldom analysed within the context of urban water issues. Water is the key component of food supply, energy generation, transport, carrier of pollutants and harmful substances, and the key to ecosystem and human health. It causes floods that on average influence some 2 billion people a year, of which a rapidly growing proportion live in cities. The world's megacities alone import an amount of virtual water equal to all the virtual water involved in the total global international food trade.

The water issues of megacities reach beyond the cities themselves. Turton *et al.* (2006, this volume) illustrate how the water requirements and the waste that Johannesburg produces influence the management of water resources on a sub-continental scale, even creating international tensions in Southern Africa. Equally, the economic wealth produced by Johannesburg has an international dimension. This feature of large urban agglomerations brings together, and even spectacularly exceeds, the paradigm of river basin management.

The same applies for the social and economic aspects. Many, if not most, megacities are global-level economic hubs that link their hinterlands, in some cases from several nations, to the globalizing economy. Thereby, financial questions have a dimension whose breadth exceeds city limits and even national boundaries. Institutional and governance challenges are of similar dimensions. Even broader is the question of the various material and energy inputs that a megacity requires. According to the earlier estimates, food imports alone to the world's megacities are of the same volume as the total international food trade. The production of such an amount of food consumes an approximate $1000 \, km^3$ of water per year. This corresponds to more than 18 times the flow of the Nile at Aswan, and the growth of the megacities is so fast that in less than five years, another Nile will be needed to feed them. This has massive implications for management of water resources on a global scale, since most of this food is produced using irrigation, and thus the hydrological cycle is modified noticeably, both in quantity and quality.

Similar examples can be found from the input of energy or, say, paper products, which depend considerably on water. Megacities are major consumers of energy and other resources, and import them in remarkable quantities. No one has estimated how large a proportion of the world's hydroelectricity is consumed in megacities but the share must be noteworthy. Bangkok accounts for one-third of Thailand's electricity consumption. Energy consumption per capita in Lagos is seven times greater than in the rest of the country. In China, urban energy consumption is about three times greater than in rural areas, and growing much faster. Tortajada (2003) estimates that in Mexico City, some 20% of electricity consumption is used to pump water and wastewater.

It is obvious that different countries have profoundly different capabilities in tackling urban problems. However, this is only partly due to economic or human capacity. An important factor is the governance dimension and the setting of political priorities. In Singapore (Tortajada 2006b, this volume), São Paulo (Braga *et al.*, 2006, this volume) and Johannesburg (Turton *et al.*, 2006, this volume), or in countries such as China or Turkey, priority has been given. An example of the reverse is Jakarta where public interest in water management seems to be low and the challenge is eluded by handing it to two foreign private operators who, as is obvious, cannot make big sustainable profits; hence development plans are not implemented (Lanti, 2006, this volume). Another negative example is Mexico City Metropolitan Area (Tortajada, 2006a, this volume), where social, environmental, economic and social policies are inadequate, as are long-term strategies for water management. The importance of strong regulatory agencies cannot be stressed enough (Lundqvist *et al.*, 2005).

Obviously the commonly accepted water management principles such as Integrated Water Resources Management (IWRM) should be developed and applied in megacities in the same way as elsewhere. IWRM is based on the so-called 3E principle: water should be used to provide economic well-being for the people, without compromising social equity and environmental sustainability. Waters should be managed in a basin-wide context, with stakeholder participation and under the prevalence of good governance.

The requirement of the basin-wide dimension is particularly difficult in this context. Many megacities are located in relatively small river basins. Many also depend on large-scale water transfer schemes that convey water from one basin to another. The material cycles of such megacities, including water and virtual water, are dominated by trade and administrative decisions rather than by hydrological boundaries. Their water sector is challenged by financial, political and institutional challenges together with economic, social and environmental ones.

The various roles of political, economic, institutional and governance-related externalities deserve far more attention in the water sector discussion than they do at present (cf. Biswas

Table 5. Some important issues of water management in megacities

Drivers	Challenges
Poverty, squatters, informal sector	*Governance*: Top-down vs. Bottom-up
Food, energy, transport and water	*Tensions*: Rural vs. Urban
Housing, real estate and water	*Bottleneck*: Conservation vs. Development
Water supply, sanitation, treatment	*Approach*: Traditional vs. Modern
Floods and droughts	*Institutions*: Informal vs. Formal
Land subsidence and groundwater	*Time priority*: Short term vs. Long term
Economy and water	*Responsibility*: Public vs. Private
Environment, wastes and water	*Management unit*: Jurisdictions vs. River basins (cf. IWRM)
Ecosystems and water	*Management philosophy*: Resource based vs. Human based
Human resources and water	

et al., 2005; Varis, 2005b). Table 5 shows some crucial issues, albeit not the only ones, that would belong to an integrated study on water management in the megacity context.

All in all, most of the growing, human-induced water challenges of the world are very closely related to urbanization, either directly or indirectly. This is due to the simple fact that virtually all population growth ends up in urban areas. Most water demand augmentation can be directly or indirectly attributed to the growing demand for food, municipal water supplies, industrial supplies, and energy, areas among others. Meeting these demands is a global concern; particularly in the case of megacities, and not only a basin-wide issue as water experts tend still to maintain.

Megacities are economical and political hubs and therefore their ability to deal with water-related problems is obviously greater than that of most other cities. However, the dimension of these cities is tremendous, posing unforeseeable problems in many ways. Not all megacities attract enough investment, nor are they economically and politically strong enough to cope in a satisfactory manner with their challenges. In fact, a large number of very poor cities in Africa as well as almost equally poor ones in Asia, particularly South Asia, are reaching megacity dimensions, and most will be unable to become global-level economic hubs in a way that many of the study cities have been able to do. In fact, the megacity problematic is still in its infancy and what we do with it now will have huge implications for a considerable portion of mankind in decades to come.

Whereas the growth of the megacities is an enormously rapid process, it must be tackled with long-term insight and strong policies. Human development, infrastructure development, economic growth, institutional development, improvements in governance and other necessities for the successful management of water problems in megacities are all outcomes of long-term evolution. Therefore, we must look far beyond the time horizon of the UN approach of talking about meeting the Millennium Goals by 2015. From short-term changes in indicators, we must go to scrutinizing structural and strategic development pathways. This does not mean that action should not be taken all the time but it means that conscious government policies that look far beyond short-term financial rationality are seriously needed.

Acknowledgements

The comments and support by Pertti Vakkilainen, Virpi Stucki, Muhammad Mizanur Rahaman and Ulla Heinonen are greatly appreciated. The feedback received at the Water and Megacities Seminar during the SIWI

World Water Week, 15 August 2004 in Stockholm is very much appreciated, in particular the cooperation with and comments of Asit K. Biswas, Jan Lundqvist and Cecilia Tortajada.

References

Abderrahman, W. (2000) Urban water management in developing arid countries, *Water Resources Development*, 16, pp. 7–20.

ADB (2004) *Bangladesh: Environmental Data* (Manila: Asian Development Bank). Available at http://www.adb.org/vehicle-emissions/BAN/environment.asp?pg=bangdesh

Ahmad, E. & Kamal, M. M. (2004) Water management in Dhaka city, in: M. Q. Hassan (Ed.) *Water Resources Management and Development in Dhaka City*, pp. 33–38 (Dhaka: Goethe-Institut).

Altinbilek, D. (2006) Water management in Istanbul, 22(2), pp. 241–253.

Baloch, L. (1999) *Poor Pay More for Water*, 7 March, UNDP/DAWN.

Bhatia, R. & Falkenmark, M. (1992) Water resource policies and the urban poor: innovative approaches and policy imperatives. Background paper for the International Conference on Water and the Environment, Dublin.

Biswas, A. K., Varis, O. & Tortajada, C. (Eds) (2005) *Integrated Water Resources Management in South and Southeast Asia* (Delhi: Oxford University Press).

Braga, B. P. F., Porto, M. F. A. & Silva, R.T. (2006) Water management in Metropolitan São Paulo, 22(2) pp. 337–352.

Cairncross, S. (1990) Water supply and the urban poor, in: J. E. Hardy, S. Cairncross & D. Satterthwaite (Eds) *The Poor Die Young: Housing and Health in Third World Cities* (London: Earthscan).

Das Gupta, A. (2005) Challenges for sustainable management of groundwater use in Bangkok, Thailand, *Water Resources Development*, 21, pp. 453–464.

ECLAC (1999) *The Impact of Macroeconomic Environment on Urban Poverty* (Santiago de Chile: Economic Commission of Latin America and the Caribbean).

Esim, S. (1996) The Gambia, in: L. Webster & P. Fidler (Eds) *The Informal Sector and Microfinance Institutions in West Africa* (Washington DC: World Bank).

Haapala, U. (2002) *Urbanisation and Water* (Espoo: Water Resources Laboratory, Helsinki University of Technology). Available at www.water.hut.fi/wr/research/glob/pubications/Haapala/alku_v.html.

Haq, K. A. (2006) Water management in Dhaka, 22(2), pp. 291–311.

Heinonen, U. (2004) The city where all streams meet: Bangkok Metropolitan Region and water management. Seminar on Water Management in Megacities, SIWI World Water Week, Stockholm, 15 August.

Hewett, P. & Montgomery, M. (2001) Poverty and public services in developing-country cities, Policy Research Division Working Paper No. 154, New York, Population Council.

Hoekstra, A. Y. & Hung, P. Q. (2002) Virtual water trade: a quantification of virtual water flows between nations in relation to international crop trade, Value of Water Research Report Series No.11 (Delft: IHE).

IFRC World Disasters Report 2002 (Geneva: International Federation of Red Cross and Red Crescent Societies). Available at www.ifrc.org/publicat/wdr2002

King, A. D. (1991) *Urbanism, Colonialism and the World Economy: Cultural and Spatial Foundations of the World Urban System* (London: Routledge).

Langman, J. (2003) The search for good jobs, *Newsweek* (22 December), pp. 40–42.

Lanti, A. (2006) Regulatory approach to the Jakarta water supply concession contracts, 22(2), pp. 255–276.

Lundqvist, J., Appasamy, P. & Nelliyat, P. (2003) Dimensions and approaches for Third World city water security, *Philosophical Transactions of the Royal Society B: Biological Sciences*, 358, pp. 1985–1996.

Lundqvist, J., Biswas, A., Tortajada, C. & Varis, O. (2005) Water management in megacities, *Ambio*, 34, pp. 269–274.

Nakicenovic, N., Grübler, A. & McDonald, A. (Eds) (1998) *Global Energy Perspectives* (Cambridge: Cambridge University Press).

Oyang, Z. & Wang, R. (2000) Water environmental problems and ecological options in China. Paper at China Water Vision: Meeting the Water Challenge in Rapid Transition. The Second World Water Forum, The Hague, 17–22 March.

Pednekar, S. S. (1997) Resource management in the Thai Mekong basin. Mekong Working Papers 71, Asia Research Centre, Murdoch University, Perth, WA. Available at wwwarc.murdoch.edu.au/arc/wp/wp71.rtf

Rahman, A., Lee, H. K. & Khan, M. A. (1997) Domestic water contamination in rapidly growing megacities of Asia: case of Karachi, Pakistan, *Environmental Monitoring and Assessment*, 44, pp. 339–360.

Rosegrant, M. W., Paisner, M. S., Mejer, S. & Witcover, J. (2001) *Global Food Projections to 2020: Emerging Trends and Alternative Futures* (Washington, DC: The International Food Policy Research Institute).

Rosegrant, M. W., Cai, X. & Cline, S. A. (2002) *World Water and Food to 2025: Dealing With Scarcity* (Washington DC: The International Food Policy Research Institute).

Söderlund, L., Sippola, J. & Kamiyo-Söderlund, M. (Eds) (2005) *Sustainable Agroecosystem Management and Development of Rural-Urban Interaction in Regions and Cities of China* (Jokioinen: MTT Agrifood Research Finland).

Todaro, M. (1997) *Economic Development*, 6th ed. (London and New York: Longman).

Tortajada, C. (2003) Water management for a megacity: Mexico City Metropolitan Area, *Ambio*, 32, pp. 124–129.

Tortajada, C. (2006a) Water management in Mexico City Metropolitan Area, 22(2), pp. 353–376.

Tortajada, C. (2006b) Water management in Singapore, 22(2), pp. 227–240.

Turton, A., Schultz, C., Buckle, H., Kgomongoe, M., Maluleke, T. & Drackner, M. (2006) Gold, scorched earth and water: the hydropolitics of Johannesburg, 22(2), pp. 313–335.

UN (2002) *World Urbanization Prospects: the 2001 Revision* (New York: United Nations).

UN-HABITAT (2003a) *Slums of the World: The Face of Urban Poverty in the New Millennium?* (Nairobi: United Nations Human Settlements Programme).

UN-HABITAT (2003b) Water and Sanitation in the World's Cities. Local Action for Global Goals (Nairobi, London: United Nations Human Settlements Programme UN-HABITAT and London: Earthscan).

UN-HABITAT (2004) *Dialogues. The Second World Urban Forum, Barcelona, 13–17 September* (Nairobi: UN-HABITAT).

UNESCO (2003) *Groundwater Resources of the World/WHYMAP*. 1st draft (Paris: UNESCO).

van Riesen, S. (1999) National Report: Germany, in: *International Report 9: The Present State of Sewage Treatment*. IWSA World Water Congress, Buenos Aires, Argentina, 20–24 September.

Varis, O. (2001) Informal water institutions. CD Proceedings of the IWA 2nd World Water Congress, Berlin, 15–19 October.

Varis, O. (2005a) Bioenergy and water—concepts, trends and magnitudes. Seminar on Water and Energy. SIWI World Water Week, Stockholm, 21 August.

Varis, O. (2005b) Externalities of integrated water resources management in South and Southeast Asia, in: A. K. Biswas, O. Varis & C. Tortajada (Eds) *Integrated Water Resources Management in South and Southeast Asia*, pp. 1–38 (Delhi: Oxford University Press).

WHO & UNICEF (2000) *Global Water Supply and Sanitation Assessment 2000 Report* (Geneva, New York: World Health Organization, UNICEF).

World Bank (2003) *World Development Indicators on CD-ROM* (Washington DC: World Bank).

Water Management in Singapore

CECILIA TORTAJADA

Introduction

Singapore is a city-state with an area of about $700 \, \text{km}^2$, a population of approximately 4.4 million people, and an annual growth of 1.9%. Total fertility rates have declined from 1.7 in 1996 to 1.4 in 2001. The population growth of Singapore between 1980 and 2005 is shown in Figure 1.

The average GDP growth of 7.7% per year during the last decade has resulted in economic prosperity, which has been translated into steady improvements in the socio–economic conditions of the country.

One of the main concerns of the government has been how to provide clean water to the population, which currently consumes about 1.36 billion litres of water per day. Singapore is considered to be a water-scarce country not because of lack of rainfall (2400 mm/year), but because of the limited amount of land area where rainfall can be stored. Singapore imports its entitlement of water from the neighbouring Johor state of Malaysia, under long-term agreements signed in 1961 and 1962 when Singapore was still a self-governing British colony. Under these agreements, Singapore can transfer water from Johor for a price of less than 1 cent per 1000 gallons until the years 2011 and 2061, respectively. The water from Johor is imported through three large pipelines across the 2 km causeway that separates the two countries.

In August 1965, Singapore became an independent country. The Constitution of Malaysia was amended on 9 August 1965. Under clause 14, this amendment stipulated that:

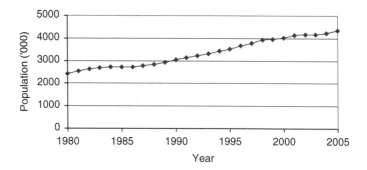

Figure 1. Total population of Singapore, 1980–2005

14. The Government of Singapore shall guarantee that the Public Utilities Board of Singapore shall on and after Singapore Day abide by the terms and conditions of the Water Agreements dated 1st September 1961, and 29th September 1962, entered into between the City Council of Singapore and the Government of the State of Johor.

The Government of Malaysia shall guarantee that the Government of the State of Johor will on and after the Singapore Day also abide by the terms and conditions of the said two Water Agreements.

The long-term security of water was an important consideration for Singapore when it became a newly independent nation. Accordingly, it made a special effort to register the Separation Agreement in the United Nations Charter Secretariat Office in June 1966.

The two countries have been negotiating the possible extension of the water agreement. The results thus far have not been encouraging since the two countries are still far apart in terms of their national requirements. Singapore would like to ensure its long-term water security by having a treaty that will provide it with the stipulated quantity of water well beyond the year 2061. In contrast, the main Malaysian demand has been for a much higher price of water, which has varied from 15 to 20 times the current price. While Singapore has said that it has no problem paying a higher price for the water it imports from Johor, its main concern has been how the price revision will be decided, and not the concept of a higher price *per se*.

Because of this continuing stalemate, Singapore has developed a new plan for increasing water security and self-sufficiency during the post 2011-period, with increasingly more efficient water management, including the formulation and implementation of new water-related policies, heavy investments in desalination and extensive reuse of wastewater, and catchment management and other similar actions.

Institutionally, Public Utilities Board (PUB) currently manages the entire water cycle of Singapore. Earlier, PUB was responsible for managing potable water, electricity and gas. On 1 April 2001, the responsibilities for sewerage and drainage were transferred to PUB from the Ministry of the Environment. This transfer allowed PUB to develop and implement a holistic policy, which included protection and expansion of water sources, stormwater management, desalination, demand management, community-driven pro-grammes, catchment management, outsourcing to private sector specific activities which are not core to its mission, and public education and awareness programmes. The country is now fully sewered to collect all wastewater, and has constructed separate drainage and sewerage systems to facilitate wastewater reuse on an extensive scale.

Overall Approach

A main reason as to why Singapore has been very successful in managing its water and wastewater is because of its concurrent emphasis on supply and demand management, wastewater and stormwater management, institutional effectiveness and creating an enabling environment, which includes a strong political will, effective legal and regulatory frameworks and an experienced and motivated workforce. The Singapore example indicates that it is unrealistic to expect the existence of an efficient water management institution in a country, in the midst of other similar mediocre management institutions, be they for energy, agriculture or industry. Water management institution in a country can only be as efficient as its management of other development sectors. The current implicit global assumption that water management institutions can be improved unilaterally when other development sectors remain somewhat inefficient is simply not a viable proposition.

Supply Management

Singapore is one of the very few countries that looks at its supply sources in their totality. In addition to importing water from Johor, it has made a determined attempt to protect its water sources (both in terms of quantity and quality on a long-term basis), expand its available sources by desalination and reuse of wastewater and stormwater (Lee & Nazarudeen, 1996), and use technological developments to increase water availability, improve water quality management and steadily lower production and management costs. At present, PUB has an in-house Centre for Advanced Water Technology, with about 50 expert staff members who provide it with the necessary research and development support.

Over the years, there has been an increasing emphasis on catchment management. Protected catchment areas are well demarcated and gazetted (Appan, 2003), and no pollution-causing activities are allowed in such protected areas. In land-scarce Singapore, protected catchment classification covers less than 5% of the area.

The Trade Effluent Regulations of 1976 promulgated the idea of partly protected catchments, where wastewater discharges to streams require prior treatment. The effluents must have an acceptable water quality that has been defined. While many other developing countries have similar requirements, the main difference is that, in Singapore, these regulations are strictly implemented. For example, when wastes from pig farms became a major source of water contamination, the Cattle Act was legislated to restrict the rearing of cattle to certain areas in the interest of public health. This also protects the water catchments from animal wastes generated from the cattle farms. At present, half of the land area of Singapore is considered to be protected and partly protected catchment. This ratio is expected to increase to two-thirds by 2009.

Desalination is becoming an important component for augmenting and diversifying available national water sources. In late 2005, the Tuas Desalination Plant, the first municipal-scale seawater desalination plant, was opened at a cost of S$200 million. Designed and constructed by a local water company, it is the first designed, built, owned and operated desalination plant in the nation. The process used is reverse osmosis and it has a capacity of 30 mgl (million gallons per day). The cost of the desalinated water during its first year of operation is S$0.78/m^3 (Lee, 2005).

Faced with the strategic issue of water security, Singapore considered the possibility of recycling wastewater (or used water) as early as the 1970s. It opted for proper treatment of its effluents, instead of discharging them to the sea. However, the first experimental recycling plant was closed in 1975 because it proved to be uneconomical and unreliable: the technology was simply not available three decades ago to make such a plant practical.

In 1998, PUB and the Ministry of the Environment formulated a reclamation study. The prototype plant, located on a site downstream of the Bedok Water Reclamation Plant, started functioning in May 2000, and produced 10 000 m^3 of water per day. The reclaimed water from this plant was monitored regularly over a period of two years, when an expert panel gave it a clean bill of health in terms of quality and reliability (Lee, 2005).

The quality of water produced by the Bedok Water Reclamation Plant was found not only to be better than the water supplied by PUB but also met the water quality standards of the Environmental Protection Agency of the United States and the World Health Organisation (PUB, 2002, in Lee, 2005).

The water supply is thus also being increased through the collection, treatment and reuse of wastewater. Investments in 2003 were of the order of S$116 million (PUB, 2003). During the period 2002–04 the amount of wastewater that was treated has increased from 1.315 to 1.369 MCM/day (Ministry of the Environment and Water Resources, 2005).

After this successful demonstration, PUB decided to collect, treat and reuse wastewater at on extensive scale, a step that very few countries have taken. At present, with a 100% sewer connection, all wastewater is collected and treated. Wastewater is reclaimed after secondary treatment by means of advanced dual-membrane and ultraviolet technologies. NEWater is used for industrial and commercial purposes, even though quality wise it is safe to drink. Since its purity is higher than tap water, it is ideal for certain types of industrial manufacturing processes, like semiconductors which require ultrapure water. It is thus economical for such plants to use NEWater since no additional treatment is necessary to improve water quality. With more industries using NEWater, water saved is being used for domestic purposes.

A small amount of NEWater (2 mgd in 2002 and 5 mgd in 2005, or about 1% of the daily consumption of the country) is blended with raw water in the reservoirs, which is then treated for domestic use. It is expected that, by 2011, Singapore will produce 65 mgd of NEWater annually, 10 mgd (2.5% of water consumption) for indirect domestic use, and 55 mgd for industrial and commercial use (PUB, 2003, in Lee, 2005).

There are currently three plants producing NEWater at Seletar, Bedok and Kranji. These plants have a total capacity of 20 mgd and will provide water to the north–eastern, eastern and northern parts of Singapore, respectively. The distribution network for NEWater includes 100 km of pipelines. PUB has recently awarded another PPP project to construct the country's largest NEWater factory at Ulu Pandan, with a capacity of 25 mgd (Khoo 2005). This plant will supply water to the western part and central business district of Singapore. Once this plant is operational, the overall production of NEWater will represent more than 10% of the total water demand per day. The overall acceptance of this recycled ultra-pure water has been high. By 2011, NEWater is expected to meet 15% of Singapore's water needs. The number of customers of NEWater, as well as some statistical information, are shown in Table 1.

Table 1. Summary of statistical information 1995–2004

	2004	2003	2002	2001	2000	1999	1998	1997	1996	1995
Employees at the end of the year	3125	3232	3333	3426	2143	2116	2163	2138	2190	2219
Customers (Number of accounts at the end of each year)										
Water	1 173 434	1 153 195	1 129 815	1 108 255	1 063 331	1 049 438	1 013 495	974 467	942 925	910 712
NEWater	51	24	–	–	–	–	–	–	–	–
Used water	1 173 462	1 153 196	1 129 792	1 108 232	–	–	–	–	–	–
Domestic Water Consumption (lpcd)	162	165	165	165	165	165	166	170	170	172
Number of accounts served per PUB employee at the end of the year	376	357	339	324	496	496	468	456	431	NA
Capital expenditure										
Water	95.8	214.8	88.0	115.6	144.0	197.4	108.7	84.0	50.5	43.7
NEWater	58.4	89.6	96.5	12.8	–	–	–	–	–	–

Notes: In 1995 PUB was restructured to be in charge of only water supply. Previously PUB handled supply of gas and electricity in addition to water. In 2001, PUB took over the drainage and sewerage departments from what was then the Ministry of the Environment.

Source: Modified from PUB, 2004 Annual Report.

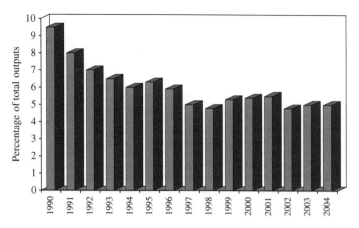

Figure 2. Unaccounted for water, Singapore, 1990–2004

The first year tender price for NEWater from the Ulu pandan plant was S$0.30/m³, which is significantly less than the cost of desalinated water. The selling price of NEWater is S$1.15/m³, which covers production, transmission and distribution costs. Because the production cost of NEWater is less than that of desalinated water, future water demands plan to be covered with more NEWater rather than with construction of desalination plants.

The supply of water is further expanded by reducing unaccounted for water (UFW), which is defined as actual water loss due to leaks, and apparent water loss arising from meter inaccuracies. Unlike other South and Southeast Asian countries, Singapore simply does not have any illegal connections to its water supply systems.

As shown in Figure 2, in 1990, unaccounted for water (UFW) was 9.5% of the total water production (Khoo, 2005). Even at this level, it would still be considered to be one of the best examples in the world at the present time. However, PUB has managed to lower the UFW consistently to around 5% in recent years. This is a level that no other country can match at present. In comparison, in England and Wales, the only region in the world which has privatized its water more than a decade ago, the best any of its private sector companies have managed to achieve is more than twice the level of Singapore. Similarly, UFW in most Asian urban centres now range between 40 and 60%.

Demand Management

Concurrent to the diversification and expansion of water sources, PUB has put in place a well-thought out and comprehensive demand management policy. It is useful to review the progress of water tariffs for water during the period 1997–2000. The progressive tariff structure used from 1997 to the present is shown in Table 2.

Before 1 July 1997, the first 20 m³ of domestic consumption for each household was charged at S$0.56/m³. The next block of 20–40 m³ was charged at S$0.80/m³. For consumption of more than 40 m³/month and non-domestic consumption, it was S$1.17/m³.

Effective from 1 July 2000, domestic consumption of up to 40 m³/month and non-domestic uses were charged at a uniform rate of S$1.17/m³. For domestic consumption of more than 40 m³/month, the tariff became S$1.40/m³, which is higher than non-domestic consumption. The earlier cheaper block rate for the first 20 m³ of domestic consumption was eliminated.

Table 2. Water Tariffs, 1997–2006

Tariff category	Consumption block (m³ per month)	Before 1 July 1997			Effective 1 July 1997		
		Tariff (¢/m³)	WCT (%)	WBF (¢/m³)	Tariff (¢/m³)	WCT (%)	WBF (¢/m³)
Domestic	1 to 20	56	0	10	73	10	15
	20 to 40	80	15	10	90	20	15
	Above 40	117	15	10	121	25	15
Non-domestic	All units	117	20	22	117	25	32
Shipping	All units	207	20	–	199	25	–

Tariff category	Consumption block (m³ per month)	Effective 1 July 1998			Effective 1 July 1999			Effective 1 July 2000		
		Tariff (¢/m³)	WCT (%)	WBF (¢/m³)	Tariff (¢/m³)	WCT (%)	WBF (¢/m³)	Tariff (¢/m³)	WCT(%)	WBF (¢/m³)
Domestic	1 to 20	87	20	20	103	25	25	117	30	30
	20 to 40	98	25	20	106	30	25	117	30	30
	Above 40	124	35	20	133	40	25	140	45	30
Non-domestic	All units	117	25	42	117	30	51	117	30	60
Shipping	All units	199	25	–	192	30	–	192	30	–

Notes: Water Conservation Tax (WCT) levied by the government to reinforce the water conservation message. Water Borne Fee (WBF) and Sanitary Appliance Fee (SAF): Statutory charges prescribed under the Statutory Appliances and Water Charges Regulations to offset the cost of treating used water and for the maintenance and extension of the public sewerage system. SAF is S$3 per sanitary fitting per month. WBF and SAF charges are inclusive of goods and services tax. *Source:* PUB (2005) personal communication.

Table 3. Average monthly consumption and bills per household, 1995, 2000, 2004

Item	1995	2000	2004
Average monthly consumption, m^3	21.7	20.5	19.3
Average monthly bill, inclusive of all taxes	S$14.50	S$31.00	S$29.40

Source: PUB (2005); personal communication.

In addition, the water conservation tax (WCT) that is levied by the government to reinforce the water conservation message, was 0% for the first 20 m^3/month consumption prior to 1 July 1997. For consumption over 20 m^3/month, WCT was set at 15%. Non-domestic users paid a WCT levy of 20%.

Effective 1 July 2000, WCT was increased to 30% of the tariff for the first 40 m^3/month for domestic consumers and all consumption for non-domestic consumers. However, domestic consumers pay 45% WCT, when their water consumption exceeds 40 m^3/month. In other words, there is now a financial disincentive for higher water consumption by the households.

Similarly, water-borne fee (WBF), a statutory charge prescribed to offset the cost of treating used water and for the maintenance and extension of the public sewerage system, was S$0.10/m^3 for all domestic consumption prior to 1 July 1997. Effective 1 July 2000, WBF was increased to S$0.30/m^3 for all domestic consumption. Impacts of these tariff increases on the consumers can be seen in Table 3.

Average monthly household consumption steadily declined during the period 1995–2004 (Table 3, Figure 3). The consumption in 2004 was 11% less than in 1995. During the same period, the average monthly bill has more than doubled.

Figure 4 shows the domestic water consumption per capita per day over the period 1995–2005. It shows a steady decline in per capita consumption because of the implementation of demand management practices, from 172 lpcd in 1995 to 160 lpcd in 2005.

These statistics indicate that the new tariffs had a notable impact on the behaviour of the consumers, and have turned out to be an effective instrument for demand management. This is a positive development since the annual water demands in Singapore increased steadily, from 403 million m^3 in 1995 to 454 million m^3 in 2000. The demand management policies introduced have resulted in the lowering of this demand, which declined to 440 million m^3 in 2004.

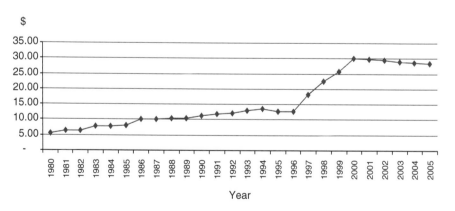

Figure 3. Average monthly bill, inclusive of all taxes (in S$) 1980–2005

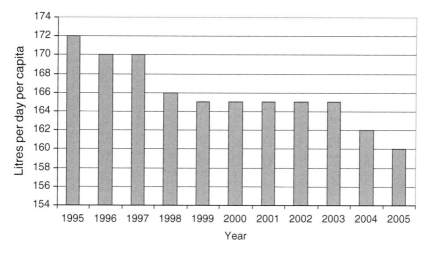

Figure 4. Domestic water consumption, 1995–2005

In terms of equity, the government provides specially targeted help for lower-income families. Households living in 1- and 2-room flats receive higher rebates during difficult economic times. For hardship cases, affected households are eligible to receive social financial assistance from the Ministry of Community Development, Youth and Sports.

The current tariff structure used by PUB has several distinct advantages, among which are the following:

- There is no 'lifeline' tariff that is used in many countries with the rational that water for the poor should be subsidized since they cannot afford to pay high tariffs for an essential requirement for human survival. The main disadvantage of such a lifeline tariff is that it also subsidizes water consumers who can afford to pay for the quantity of water they actually consume.
- The poor who cannot afford to pay for the current water tariffs receive a targeted subsidy. This is a much more efficient policy in socio-economic terms, instead of providing subsidized water to all for the first $20–30\,m^3$ of water consumed by all households, irrespective of their economic conditions.
- The current domestic tariff of water consumption up to $40\,m^3$/month/household is identical to the non-domestic tariff. Both are set at S\$1.17/m^3. In other words, commercial and industrial users do not subsidize domestic users, which is often the case for numerous countries.
- The tariff structure penalizes all those households who use more than $40\,m^3$ of water per month. They pay the highest rates, S\$1.40 m^3, for consumption above this level. This rate is higher than the commercial and the industrial rates, and is a somewhat unusual feature compared to the existing norm.
- Water conservation tax (WCT) is 30% of the tariff for all consumers, except for domestic households who use more than $40\,m^3$/month. The WCT on consumption of each unit higher than $40\,m^3$/month increases by 50%, from 30% to 45%, which must be having perceptible impacts on household behaviour in terms of water conservation and overall demand management.

- Water-borne fee (WBF) is used to offset the cost of treating wastewater and for the maintenance and extension of the public sewerage system. It is set at S\$0.30 m^3/s for all domestic consumption. For non-domestic consumption, this fee is doubled, S\$0.60/m^3, presumably because it is more difficult and expensive to treat non-domestic wastewater.
- A Sanitary Appliance Fee (SAF) is also levied per sanitary fitting per month. It is currently set at S\$3.00 per fitting.
- There are two components to water tariff. A major component of the overall revenue collected through water tariffs accrue to the PUB recovering all operation and for considering maintenance costs and new investments. However, revenue from WCT accrues to the government and not to PUB.

Overall Governance

The overall governance of the water supply and wastewater management systems in Singapore is exemplary in terms of its performance, transparency and accountability. There is much that both the developed and developing world can learn from the PUB experience. Only some selected critical issues will be discussed herein.

Human Resources

An institution can only be as efficient as its management and the staff that work for it, and the overall social, political and legal environment within which it operates. In terms of human resources, PUB has some unique features in terms of management that makes it stand out among its other Asian counterparts.

In the vast majority of the Asian water utilities, service providers mostly have a limited say on staff recruitment and staff remuneration. Consequently, the utilities are rife with following types of problems:

- Staff, including senior managers, are often selected because of their political connections, rather than their management abilities or technical skills.
- Managers often do not have the skill to manage, even if they had autonomy and authority to manage, which often they do not.
- Water utilities are overstaffed, primarily because of political interference and nepotism. Unions are very strong and generally well-connected politically. Accordingly, downsizing is a difficult task because of strong union opposition and explicit or implicit political support. Overstaffing ensures low productivity and low staff morale.
- Utilities are not allowed to pay their professional staff members the going market rates for remuneration, which sometimes could be 2–3 times higher. This means that they are unable to attract and retain right calibre of staff. Many staff moonlight to obtain extra income, and corruption is rife in nearly all levels.
- Utilities are dominated by engineers, and the career structure available for other disciplines like accountants, administrators, social scientists, information technologist, etc. is somewhat limited. This is another disincentive for non-engineers to join.
- Poor management, overstaffing and promotions because of seniority or political connections ensure that it is very difficult to recruit good staff, and if some do

join, it is equally difficult to retain them because of lack of job satisfaction, poor working environment and absence of incentives for good performance.

PUB has overcome the above and other related constraints through a competitive remuneration, incentives and benefits package. The salary and benefit package is generally benchmarked against the Civil Service, which, in turn, benchmarks against the prevailing market. It provides strong performance incentives that are commensurate with the prevailing pay packages for the private sector. In addition, its pro-family policies, commitment to train its staff for their professional and personal development, and rewarding good performers, ensure good organizational performance and development. Consequently, its overall performance has become undoubtedly one of the best in the world.

Corruption

Corruption is endemic in most Asian utilities. However, it is not an issue at PUB, which emphasizes staff integrity as a key organizational requirement. It has taken measures to prevent corruption by staff training on Code of Governance and Code of Conduct, effective internal control processes, regular audits and strong and immediate sanctions against those who may prove to be corrupt. Staff members are required to make annual declarations, which include Declaration of Assets and Investments and Declaration of Non-indebtedness.

Complaints of corruption are promptly investigated and reported to Singapore's Corrupt Practices Investigation Bureau. PUB is a part of the overall Singapore milieu where there are strong anti-corruption laws at the national level with appropriate sanctions that are regularly implemented. In addition, in recent decades, the government has consistently shown its strong political will to curb all forms of corruption, and take firm actions against all and any form of corruption (see http://www.cpib.org.sg/aboutus.htm).

With a good remuneration package, functional institution, and a strong anti-corruption culture, corruption is not an issue at PUB.

Autonomy

Absence of autonomy is one of the most fundamental problems that affect most utilities of the Asian developing countries. This, in turn, creates a series of second order problems and constraints that further erode the efficiency of the utilities to perform their tasks efficiently and in a timely manner.

A fundamental problem in most Asian cities has been that the process of setting tariffs is primarily controlled by the elected officials, who mostly resist increases because of perceived vested interests. Low levels of tariffs cannot have any impact in terms of managing demands. In fact, low levels of tariffs are not compatible with metering, especially as the cost of metering and processing the resulting information may be higher than the revenue metering can generate. The problem is further accentuated by low levels of tariff collection. Furthermore, politicians have preferred to keep domestic water prices artificially low, and subsidize it with much higher tariffs from commercial and industrial consumers. For example, according to a World Bank study, in India, domestic consumers used 90% of the water, but accounted for only 20% of the revenues (ADB, 2003). Domestic consumers were thus heavily cross-subsidized by commercial and industrial water users.

In contrast, PUB has a high level of autonomy and solid political and public support, which have allowed it to increase water tariffs in progressive steps between 1997 and 2000 (see Table 1). This increase not only has reduced the average monthly household water demand but also has increased the income of PUB, which has enabled it to generate funds not only for good and timely operation and maintenance of the existing system but also for investments for future activities. Water tariffs have not been raised since July 2000.

Such an approach has enabled PUB to fund its new capex investments over the years from its own income and internal reserves. In 2005, for the first time, PUB tapped the commercial market for S$400 million bond issue. Under the Public Utilities Act, the responsible Minister for the Environment and Water Resources had to approve the borrowing. The budgeted capex for the year 2005 was nearly S$200 million.

Because of lack of autonomy, political interferences, and other associated reasons, internal cash generation of water utilities in developing countries to finance water supply and sanitation has steadily declined: from 34% in 1988, to 10% in 1991 and only 8% in 1998. Thus, the overall situation has been 'lose–lose' for all the activities. The Singapore experience indicates that given autonomy and other appropriate enabling environmental conditions, the utilities are able to be not only financially viable but also perform their tasks efficiently.

Unlike many other similar Asian utilities, the PUB has extensively used the private sector where it did not have special competence or competitive advantage in order to strive for the lowest cost alternative. Earlier, the use of the private sector for desalination and wastewater reclamation was noted. In addition, specific activities are often outsourced to private sector companies. According to the Asian Development Bank (November 2005), some S$2.7 billion of water-related activities were outsourced over the 'last four years', and another S$900 million will be outsourced during 'the next two years' to improve the water services.

Overall Performance

No matter which performance indicators are used, PUB invariably appears in the top 5% of all the urban water utilities of the world in terms of its performance. Only a few of these indicators will be noted below:

- 100% of population have access to drinking water and sanitation.
- The entire water supply system, from water works to consumers, is 100% metered.
- Unaccounted for water as a percentage of total production was 5.18% in 2004.
- The number of accounts served per PUB employee was 376 in 2004 (Figure 5).
- Monthly bill collection efficiency: 99% in 2004.
- Monthly bill collection in terms of days of sales outstanding was 35 days in 2004.

The above analysis indicates that PUB has initiated numerous innovative approaches to manage the total water cycle in Singapore. Many of these approaches can be adopted by developed and developing countries to improve their water management systems. If the MDGs that relate to water are to be reached, the example of Singapore needs to be seriously considered for adoption by developing countries concerned and the donor community, after appropriate modifications.

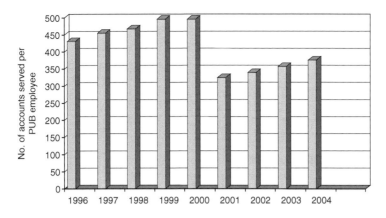

Figure 5. Number of accounts served per PUB employee, 1996–2004

Concluding Remarks

Viewed from any perspective, any objective analysis has to conclude that water supply and wastewater management practices in recent years in Singapore have been exemplary. Water demand management practices are unquestionably one of the best, if not the best, from any developed or developing countries, irrespective of whether a public or private sector institution is managing the water services. Singapore has successfully managed to find the right balances between:

- water quantity and water quality considerations;
- water supply and water demand management;
- public sector and private sector participation;
- efficiency and equity considerations;
- strategic national interest and economic efficiency; and
- strengthening internal capacities and reliance on external sources.

In other words, the country has successfully implemented what most water professionals have been preaching in recent years.

By ensuring efficient use of its limited water resources through economic instruments, adopting the latest technological development to produce 'new' sources of water, enhancing storage capacities by proper catchment management, practicing water conservation measures, and ensuring concurrent consideration of social, economic and environmental factors, Singapore has reached a level of holistic water management that other urban centres will do well to emulate.

Acknowledgements

This paper is an extended version of the one prepared for the 2006 Human Development Report. Grateful acknowledgement is made to the Public Utilities Board of Singapore, especially to its Chief Executive Officer Mr Teng Chye Khoo, for providing up-to-date information on which some of the analyses are based.

References

Appan, A. (2003) A total approach to water quality management in partly-protected catchments. Paper presented at The Singapore Experience, International Workshop on Management and Conservation of Urban Lakes, Hyderabad, India, 16–18 June.

Asian Development Bank (2003) *Asian Water Supplies: Reaching the Poor* (Manila: ADB).

Asian Development Bank (2005) Country Water Action: Singapore. Technology turns Water Weakness to Strength, November.

Cattle Act, Republic of Singapore. Available at http://app.nea.gov.sg/cms/htdocs/category_sub.asp?cid=181

Khoo Teng Chye (2005) Water resources management in Singapore. Paper presented at Second Asian Water Forum, Bali, Indonesia, 29 August–3 September.

Lee, M. F. & Nazarudeen, H. (1996) Collection of urban storm water for potable water supply in Singapore, *Water Quality International*, May/June, pp. 36–40.

Lee Poh Onn, (2005) Water management issues in Singapore, paper presented at Water in Mainland Southeast Asia, 29 November-2 December, Siem Reap Cambodia, conference organized by the International Institute for Asian Studies (IIAS), Netherlands, and the Centre for Khmer Studies (CKS), Cambodia.

Malaysia, Act of Parliament No. 53 of 1965, Constitution of Malaysia (Singapore Amendment) Act, 1965: An Act to amend the Constitution of Malaysia and the Malaysia Act, 9 August, 1965.

Ministry of the Environment and Water Resources (2006) Key Environmental Statistics 2005, Singapore.

Public Utilities Board (2002) Annual Report, Singapore.

Public Utilities Board (2003) Annual Report, Singapore.

Public Utilities Board (2004) Annual Report, Singapore.

The Johore River Water Agreement between the Johore State Government and City Council of Singapore signed on 29 September 1962. Available at http://www.channelnewsasia.com/cna/ Parliament/relations/documents.htm

The Tebrau and Soudai Rivers Agreement between the Government of the State of Johore and the City Council of the State of Singapore signed on 1 September 1962. Available at http://www.channelnewsasia.com/cna/parliament/relations/documents.htm

Tortajada, C. (2006) Singapore: an exemplary case for urban water management, prepared for the 2006 Human Development Report.

Trade Effluents Regulations (1976) Water Pollution Control and Drainage Act, Act 29/75, Pub. No. SLS 29/76, Republic of Singapore.

Water Management in Istanbul

DOGAN ALTINBILEK

Historical Perspective

Istanbul, the largest industrial, trading, tourism and cultural centre of Turkey, is a megacity with a population of 11.5 million. Istanbul is located on both the European and Asian continents, divided by the Bosphorus Strait. The coastal location of Istanbul has its advantages and disadvantages in terms of water supply and wastewater disposal.

The historical city was first established during the 7th century when it was known as Byzantine (ISKI, 2004). During the Roman rule the city was named Constantinople after Constantinus I, who built new city walls. After the Roman Empire's division into East and West, the city became the capital of the East Roman (Byzantine) Empire. Istanbul became the capital of the Ottoman Empire in 1453. The population of Istanbul was estimated to be around 100 000 during the reign of Sultan Mehmet the Conqueror. Currently, one-sixth of Turkey's population and approximately 40% of Turkish industry are located in Istanbul. Annually, 400 000 people are added to the cities' population. Such a rapid population increase creates many infrastructural problems, including the water supply sector (Eroglu et al., 1998, 2001; Codur, 2004).

Water Management Organizations

The management of the water supply to Istanbul can be grouped into five eras chronologically as follows (ISKI, 2004).

Water supply systems before the Ottomans.   Water requirements of the city were initially met by groundwater sources, small springs and cisterns. During the Roman period four long gravity transmission lines were built. Aqueducts were used to cross valleys. The longest of these transmission lines (242 km) was built by Emperor Constantinus in 324 AD. Although water supply systems built in the Roman period were repaired and extended by Byzantine Emperors, they were almost in ruins during the final period of the Byzantine Empire because of wars and sieges. More than 70 covered cisterns with a total storage capacity of 200 000 m^3 and open cisterns of 800 000 m^3 were built at this time.

Ottoman Period.   After the conquest by the Ottomans in 1453 a search and construction activity began for an additional water supply. Five new major transmission lines were built by different Sultans. The largest of these, the Kırkcesme water supply system (55 km long, 17 443 m^3/day capacity, 94 fountains, 19 wells, 15 water tanks, 13 public Turkish baths, 4200 m^3 of water storage) was built in 1563 and is still operational. Eight historical water supply dams with a total storage of 1.7 million m^3 were built. In Istanbul, there were more than 1000 historical public fountains. Diverse sources from different springs were distributed with small supply lines to public fountains. These individual water systems were developed by Sultans or other patrons as water foundations to meet the water demands of the people. The most important of these were the Hamidiye waters, constructed in 1904 that were distributed to 50 fountains with a daily volume of 1200 m^3.

Water companies.   Responsibility of providing a water supply to the European part of Istanbul was given as a franchise to a private French company called Dersaadet Water Company in 1869 to bring water from Lake Terkos. The first water treatment plant in Istanbul was constructed by this company in 1926. Another French owned company, Uskudar-Kadikoy Water Company, was founded in 1888 to meet the needs of the Asian part of Istanbul. Elmalı-1 Dam, water treatment plants and distribution networks were constructed by water companies.

Istanbul Water Administration.   After the foundation of Turkish Republic, it was believed that the problem of water in Istanbul could not be solved by franchise private companies. The operating rights of the private companies were transferred to Istanbul Water Administration in 1933 to distribute and manage water resources and to collect fees. Elmalı-2 Dam was built. The pumps were modernized and the capacities of seven transmission lines and water systems were increased. State Hydraulic Works (DSI), founded in 1954, was also active in the construction of the Omerli water supply dam and water transmission lines. The Water Supply Master Plan of Istanbul was prepared in 1971 by DSI (DAMOC, 1971).

Istanbul Water and Sewerage Administration (ISKI). In 1981, water and wastewater services were combined under the Istanbul Water and Sewerage Administration, its abbreviated name in Turkish being ISKI. The responsibilities of ISKI are the planning, construction and the operation of:

- Water supply and distribution.
- Wastewater collection, treatment and discharge.
- Protection of the water resources against pollution.
- Rehabilitation of water courses within the metropolitan area.

The Istanbul Water Supply, Sewage and Drainage Master Plan and Feasibility Report was prepared by the Istanbul Master Plan Consortium (IMC) for ISKI in 1994 (ISKI, 1994).

At present (2005), ISKI has jurisdiction over an area of 6504 km^2. The ISKI workforce consists of 2245 officers and 3765 workers, totalling 6010 personnel. The total annual operating budget of ISKI in 2005 was US$1.2 billion, including an annual investment of US$709 million, personnel expenditure of US$240 million and energy expenses of US$92 million.

The General Directorate of State Hydraulic Works (DSI) was also active in this era to supply municipal and industrial water to Istanbul. As such, DSI has undertaken several planning studies and has constructed many dams, transmission lines and treatment plants that supply an annual total of 783 million m^3 of water, corresponding to two-thirds of the available total supply capacity at present (DAMOC, 1971). DSI is also executing another very large water supply project, namely Greater Melen System which will meet the water needs of Istanbul until the year 2040 (DSI, 2000).

Population and Water Demand

The current population of the Istanbul metropolitan area is estimated at approximately 11.5 million. The metropolitan (city) population accounted for 91% of the total population of Istanbul Province in 2000. Population growth in the city is almost twice the overall rate for the whole of Turkey because of large in-migration. The change of population in the Istanbul metropolitan area from 1955, when the first official census was conducted, is shown in Figure 1. The population of the city has experienced an average annual growth rate of 4.5% over the last half a century. In the five-year period from 1980–85, the city

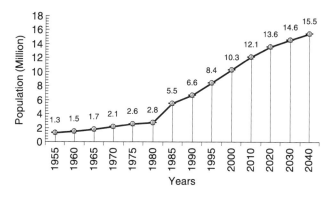

Figure 1. Population of Istanbul

grew 14.6% annually. Istanbul experienced severe water shortages in the early 1990s. A population of around 400 000 annually migrates to Istanbul. It could be said that Istanbul's growth is equal to that of a medium sized city in Europe. For this reason, the projections for the demand for water were underestimated. Thus Istanbul's problems such as potable water, sewerage and treatment of wastewater remained unsolved for many years. The Master Plan prepared in 2004 has 2040 as the target year (ISKI, 1994).

The projected water demand and supply of Istanbul is shown in Figure 2 and Table 1. To determine the water demands of Istanbul, the domestic water need is calculated by taking the daily individual consumption as 250 litres. The total annual water need of Istanbul in 2004 was 709 million m^3 per year. The water demand in 2020 is estimated to be 1059 million m^3 per year.

Water Supply System

The available water resources of Istanbul are shown in Table 2 and a schematic of the water supply projects is shown in Figure 3. The available water resources developed on the European side are 648 million m^3/year and 522 million m^3/year on the Asian side. At present, two-thirds of the population lives on the European side and one-third lives on the Asian side. Contrary to the population distribution, potentially more water is available on the Asian side. To overcome the geographical misdistribution, water is transported under the sea from the Asian side to the European side by two pipelines, each 1 m diameter. At present 126 million m^3 water is being transferred from the Asian side to the European side of the city. In the future a 4.0 m diameter tunnel will be constructed to supply the European side of the city from the Greater Melen Project in the Asian part of the city.

The historical development of the water supply projects is shown in Table 3. During the last 10 years an additional capacity of 595 million m^3/year was developed by new projects of ISKI and DSI. The total investment made for water systems during the period 1994–2004 was US\$ 3.6 billion.

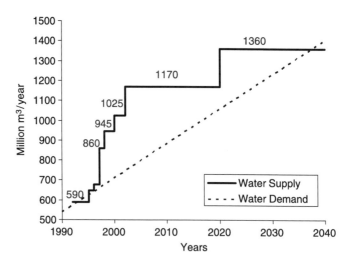

Figure 2. Water supply and demand

Table 1. Projected water demand

Year	Population (million)	Water demand (million m³/year)	Water demand (lt/cap/day)	Water demand (million m³/day)
1990	6.60	511	152	1.4
1995	8.40	657	214	1.8
2000	10.30	694	184	1.9
2010	12.10	913	206	2.5
2020	13.60	1059	212	2.9
2030	14.60	1241	232	3.4
2040	15.50	1387	245	3.8

The characteristics of Istanbul's water and wastewater system for 1994 and 2004 are shown in Table 4 (ISKI, *Annual Reports*). The capacity of the water supply sources is currently 1170 million m³ per year, while the average annual water supply was 709 million m³ per year in 2004. Istanbul's water supply is provided from 15 different dams, one natural lake and 82 groundwater wells. The raw water received from the dams is treated at five different treatment plants with a daily capacity of 3.5 million m³. Fifty-one service reservoirs with a capacity of 952 000 m³ and 58 pumping stations with an installed capacity of 273 084 kW serve a distribution network of 12 850 km of ductile iron pipes.

Large-scale Water Transmission Projects for Istanbul

The 1971 Master Plan for the Istanbul water supply suggested that 11 water sources could be developed for Istanbul (DAMOC, 1971). Six of these have already been put into service. However, the remaining five could not be implemented due to geological problems, unsuitable water quality or because of low capacities involved. The General Directorate of State Hydraulic Works (DSI) has begun two large projects in order to meet the water needs of Istanbul in the medium and long term, namely the Yesilcay project and Greater Melen project.

The Yesilcay System (Figure 4) transmits 145 million m³ of water per year to Istanbul from a distance of 60 km along a 3 m diameter pipeline. The Yesilcay project, at a cost of US$ 271 million, will supply drinking water and the domestic needs of an additional

Table 2. Available water resources in Istanbul (2005)

European side	10⁶ × m³ per year	Asian side	10⁶ × m³ per year
Terkos Lake (1883)[a]	162	Omerli Dam (1972)[a]	235
B. Cekmece Dam (1989)[a]	120	Darlik Dam (1989)	97
Sazlidere Dam (1998)[a]	85	Yesilcay Project (2003)[a]	145
Alibeykoy Dam (1972)[a]	36	Elmali I&II Dams (1893–1950)	15
Istranca Project (1997–2000)	235	Others	30
Others	10		
Total	648	Total	522

[a] Projects that were developed by State Hydraulic Works (DSI).

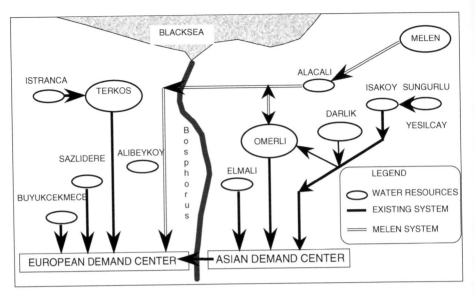

Figure 3. Schematic of water supply projects

population of approximately 1.5 million. The Yesilcay project became operational in May 2003.

The Greater Melen System (Figure 5) will supply 268 million m^3 of water per year to Istanbul is in its first stage (DSI, 2000). At the end of the fourth stage, the system will provide 1.18 billion m^3 water per year to Istanbul to meet demands until the year 2040. The first stage of the Greater Melen project involves 180 km of 2.5 m diameter pipeline, 25 km tunnels and a water treatment capacity of 3.5 m^3 per day. The drinking and domestic water demands of an additional 2.75 million people will be supplied. The total cost of the first stage of Greater Melen System is US$1.18 billion.

In addition, from 1995–2000 ISKI developed the Instranca water project from various streams on the Black Sea coast on the European side of the city. Seven dams were built to transfer water (235 million m^3 per year) 41.3 km with a 1.8 m diameter pipeline.

Wastewater Collection, Treatment and Disposal

During the Ottoman period, wastewater and stormwater systems were constructed perpendicular to the coast. Since the discharges involved were small, pollution in the

Table 3. The historical development of water supply system of Istanbul

Period	Capacity added (million m^3/year)	Water supply systems
1884–1974	413	Elmali 1–2, Terkos, Alibeykoy, Omerli
1974–94	162	Darlik, B. Cekmece, Yesil Vadi
1994–2005	595	Duzdere, Kuzuludere, Buyukdere, Sile, Elmalidere, S. Bahcedere, Kazandere, Sazlidere, Pabucdere, Yesilcay

Table 4. Characteristics of Istanbul's water and wastewater system

Characteristics	1994	2004
Population served (million)	8.1	11.5
Total area served (km²)	1976	6504
Total number of customers	1 715 138	3 528 556
Length of distribution network (km of ductile iron)	5957	12 815
Lenght of transmission lines (km)	226.7	851
Length of sewer canals (km)	9005	9602
Capacity of the water supply sources (million m³/year)	590	1170
Average daily water supply (m³/day)	1 074 000	1 942 000
Capacity of water treatment plants (m³/day)	1 078 000	3 558 000
Volume of the water reservoirs (1000 × m³)	163	805
Pumping stations (installed power in kw)	59 000	241 011

receiving water bodies was negligible. With a rapid increase in population, the construction of wastewater collection systems became a necessity (Eroglu *et al.*, 1998, 2001).

At present, the wastewater collection system consists of 1868 km of wastewater network, 263 km collectors, 284 km main collectors, 50 km tunnels and 17 pumping stations. The wastewater treatment system consists of seven pre-treatment plants and five biological treatment plants. The percentage of wastewater treated and disposed of is shown in Table 5. Since 1995 there has been a sharp increase in the percentages of wastewater treated. The present capacity of the wastewater treatment plants is 3.9 million m³ per day. Wastewater treatment capability consists of tertiary (8.3%), secondary (28.2%) and mostly preliminary treatments. However, there are plans to add new tertiary treatment facilities exceeding one million m³ per day capacity.

Treated wastewaters are discharged into the Bosphorus Strait that joins the Black Sea and the Marmara Sea. Studies and surveys have demonstrated that about 90% of the discharges into the lower layer of the two phase flow along the Bosphorus Strait reach

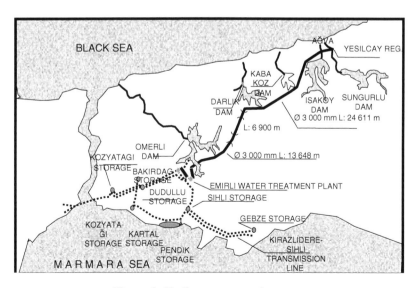

Figure 4. Yesilcay water supply system

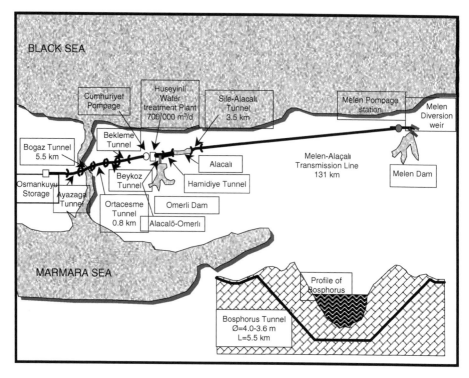

Figure 5. Greater Melen water supply system

the Black Sea. Ecological modelling studies have shown that primary treatment is sufficient for sea disposal directly into the Black Sea while tertiary treatment is required for disposal into the Marmara Sea (Eroglu *et al.*, 1998). A total of 13.9 km of land pipes and 11.5 km sea pipelines are used for sea disposal units.

Table 5. The percentage and amount of the treated wastewater

Year	Treated wastewater	
	Percentage	Amount (1000 m^3)
1993	9.3	192
1994	10	208
1995	11	211
1996	16	214
1997	47	370
1998	63	556
1999	70	627
2000	80	790
2001	90	865
2002	95	942
2003	95	1370
2004	95	1755

Water Quality

A large percentage (97%) of Istanbul's water demand is met from surface sources such as dams, lakes and diversion weirs. The rest is supplied from groundwater and historical small dams. The processes commonly applied to surface water treatment at five different locations are: preoxidation (chlorination or ozonation), coagulation, flocculation, sedimentation, filtration and post-disinfection. Powdered activated carbon absorption is applied only at one treatment plant. Water quality in the network is monitored and checked by four different metropolitan agencies. On average 350 to 400 samples are collected daily from sampling points of the distribution network and analyzed for chemical and bacteriological parameters. The measurements yield the results surpassing the quality specified by the Turkish Standards Institute (TSE 226), World Health Organisation, US Environment Protection Agency and European Community. With the completion of new treatment plants and the water distribution network, water quality has been improved. In a recent survey in 2004, 35% of the customers stated that they drink water from the taps, which is an increase from 10% in 2000.

Unaccounted-for Water

The unaccounted-for water in the parts of Istanbul's water distribution network was more than 50% prior to 1994. This loss was largely due to the age of the water distribution system and lack of proper maintenance. Large investments have been made since 1994 to improve the system. As a result, the unaccounted-for water dropped to 34% in 2000. By definition, unaccounted-for water includes the public uses such as baths, mosques, cemeteries, fire hydrants and as well as losses due to pipe breaks and repairs. After implementation of the pipe replacement programme, the target to be achieved for unaccounted-for water is 25%.

Water Tariffs

Unlike the previous agencies it has replaced, ISKI has the right to set the water and sewerage tariffs without outside approval. Different rates are charged for domestic, commercial and industrial consumption in Istanbul. Sewer charges are included in the water tariffs. The water rates effective in January 2005 are shown in Table 6. Water tariffs are set to cover not only the operation and maintenance expenses but also to create financial resources for future investments.

About 70% of Istanbul's population is regarded as low-income. Water tariffs are highly progressive with industrial and commercial rates higher than residential rates, involving

Table 6. Water rates, US\$/m^3 (January 2005)

Residences-1 (0–25 m^3/day)	1.10
Residences-2 (>26 m^3/day)	2.20
Commercial and industrial uses	2.20
Schools	0.95
Hospitals and public sector	1.84

substantial cross-subsidization. Consumers using less than $25\,m^3$ a month pay half the price compared to those who use more. Tariffs for households below the poverty line result in charges of less than 3% of household income and seem to be well accepted by consumers. Residential consumers have a better payment record than industries and public institutions, which have particularly poor payment records.

The Environmental Protection of Watersheds

The water supply of Istanbul is largely met by surface water, various dams and reservoirs. It is of utmost importance that the watersheds of these reservoirs are protected against settlement and pollution. Unfortunately, during recent decades the watersheds have come under ever increasing pressure from the population. Watersheds have been opened up for both residential and industrial construction. For the protection of watersheds, several measures must be put into effect for a sustainable water supply (Akkoyunlu *et al.*, 2002; Codur, 2004).

Watershed Protection Zones

ISKI has set the protection zones for the environmental control of the water reservoirs. The banks of every water supply reservoir are protected by an Absolute Protection Zone (0–300 m), a Proximate Protection Zone (300–1000 m), a Mediate Protection Zone (1000–2000 m) and a Remote Protection Zone (2000–5000 m), with various restrictions for each zone.

Within the Absolute Protection Zone, no settlement is allowed. Existing buildings are expropriated and demolished. Agricultural and mining activities are not permitted. Use of fertilizers and pesticides is not allowed. Forestation is programmed.

In the Proximate Protection Zone surrounding a drinking water supply reservoir, a residence permit is granted for two-storey buildings to allow a population density of less than five persons per hectare of land. The infrastructure developed is required not to pollute the reservoir. Industrial development, mining, cemeteries, agricultural and animal farming are not allowed.

In the Mediate Protection Zone, two-storey residences are allowed with a density not exceeding 10 persons per hectare of land. In this area, the use of chemical fertilizers and pesticides is not allowed. Industrial, warehouse and hospital buildings as well as mines, cemeteries and deposition of solid wastes are not allowed.

In the Remote Protection Zone, two-storey residences are allowed, not exceeding a population density of 20 persons per hectare of land. Again, no permits are granted for industrial use, hospitals, chemical storage and mining facilities as well as garbage disposal and cemeteries.

Creeks

ISKI has defined Absolute Protection Zones with a width of 100 m on both banks of creeks flowing into the reservoirs in the watersheds. Any building within these bands is acquired and demolished to control pollution as well as to allow the unobstructed flow during floods (Eroglu *et al.*, 2002)

Administrative Precautions

Special teams have been formed by ISKI to guard the watersheds against unauthorized construction activities. Emergency Action Teams under the Istanbul Governorship are also active to take action in the case of unauthorized construction activities. In the last 10 years 1208 such buildings were demolished (Codur, 2004).

Technical Precautions

To protect watersheds, forestation activities are underway. In the last 10 years, 408 750 trees have been planted. Land acquisition of the Absolute Protection Band (0–300 m) of reservoirs is continuing. The stream channels in the watersheds are rehabilitated and those riverbeds that are obstructed due to occupation are emptied. In order to protect the reservoirs, wastewater collection network investments are accelerated.

Use of Satellites

A remote sensing capability was set up to inspect watersheds from satellite images. ISKI utilizes Turksat data to find unlicensed constructions and to trace pollution. Once an unauthorized construction is detected, the Municipal Teams demolish the structure, which then discourages further construction. The identification of topological structural change by remote sensing is effective for the global assessment of geometric characteristics and an appraisal of land-use analysis.

Use of GIS Technology

ISKI utilizes GIS technology to trace the water quality in the watersheds, at treatment plants, at transmission mains and service reservoirs and within the distribution network. A website (www.iski.gov.tr) is being developed in order to present data to users on the Internet.

Golden Horn Environmental Rehabilitation Project

Istanbul's rapid growth has led to large squatter settlements with inadequate and ill-maintained water supplies and wastewater. The population, estimated at 2.8 million in 1980, increased to 8.4 million in 1995. Wastewaters were discharged untreated into the Golden Horn, the Sea of Marmara and the Bosphorus. Over time, the static parts of the waters of the Golden Horn, which is an estuary connected to the Marmara Sea, became anaerobic and the beaches of the Sea of Marmara became highly contaminated. In 1985, there were 700 industrial sites and 2000 businesses on the banks of the Golden Horn producing a smell that caused great discomfort to the people living in the area. ISKI began an environmental rehabilitation project in 1994 that involved:

- Blocking and treating incoming waters to the Golden Horn at treatment plants.
- Dredging the sludge, totalling approximately 5 million m^3.
- Removing the Galata Bridge in order to sustain currents.
- Closing the industrial sites and dockyards on the banks.
- Constructing recreation areas, green fields and parks on the banks.

This rehabilitation project, which has been one of the world's largest estuary cleaning projects, has cost a total of US$653 million. While once the biological life in the water of the Golden Horn was extinct, 33 differences species of fishes are now observed in the estuary. The value of the coliform bacteria per 100 millilitres is 1000, while it was as much as 350 000 in 1998. The estuary is now open to swimming, rowing and navigation. The Golden Horn Environmental Protection Project was awarded First Prize in the Metropolis Award in 2002.

Conclusions

From this case study of the Istanbul water supply and management, the following conclusions can be deduced with regard to the water problems faced by Istanbul and other similar megacities in the developing world.

- Water supply emerged as a critical issue in late 20th century. With the rapid population increase due to urbanization and high natural growth as observed in Istanbul, the effective implementation of water supply and wastewater collection systems has become an urgent necessity. In a world where available fresh surface water resources are unevenly distributed with regard to time and space, meeting the needs of the megacities require huge investments.
- The establishment of an independent Metropolitan Water Agency such as ISKI is useful for the effective management of the water systems.
- The pricing of water is one of the most important factors that will allow the Metropolitan Water Agencies such as ISKI to have a sufficient budget for future investments. In cases of the inability to pay by customers, governments must use tax revenues either to enable private companies to undertake the investments or to make large investments directly to avoid future shortages of supply.
- When the construction of water supply investments takes precedence over wastewater collection and treatment investments, environmental pollution of various scales will be experienced.
- The introduction of up-to-date techniques such as Information Technologies, SCADA, remote sensing and GIS can be useful for the effective operation and protection of the systems.
- In addition to infrastructure investments on the water supply side, the reduction of unaccounted-for water in the distribution system and other water saving strategies must be considered.
- The accelerated urbanization as experienced in Istanbul can endanger the watersheds of the water supply reservoirs, jeopardizing the existing infrastructure in short periods. Necessary technical and administrative precautions must be taken in time to prevent environmental pollution.

Acknowledgements

This paper was originally submitted at a seminar at the XI World Water Congress, which took place in Madrid in October 2003. It was also presented at a seminar during the Stockholm Water Symposium and World Water Week in August 2004. The author would like to express his sincere appreciation and gratitude to Prof. Asit K. Biswas and Dr. Cecilia Tortajada of the Third World Centre for Water Management who organized and chaired both seminars. Special thanks are due to Prof. Dr. Veysel Eroglu, Director General of DSI and Past Director General of

ISKI and also to Mr. Dursun Ali Codur, Director General of ISKI for generously providing the data required to bring about and to update this work.

References

Akkoyunlu, A., Yuksel, E., Erturk, F. & Bayhan, H. (2002) Managing of watersheds of Istanbul (Turkey), Paper presented at the *Fifth Water Information Summit: Regional Perspectives on Water Information Management Systems* (Fort Lauderdale, FL: Water Web Consortium).

Codur, D. A. (2004) Istanbul's water supply, water quality and watershed protection, *Proceedings of Symposium on Istanbul and Water, Organized by the Turkish Chamber of Architects*, Istanbul Branch pp. 142–157 (Istanbul) (in Turkish).

DAMOC Consortium (1971) *Master Plan and Feasibility Reports for Water and Sewerage for the Istanbul Region* (Istanbul: DAMOC).

Eroglu, V. & Sarıkaya, H. Z. (1998) Achievements towards better water supply and wastewater disposal in Istanbul, in: *Proceedings of an International Symposium on Water Supply and Treatment, Organized by Istanbul Water and Sewerage Administration*, pp. 1–19, Istanbul, 25–26 May.

Eroglu, V., Sarikaya, H. Z. & Aydin, A. F. (2001) Planning of wastewater treatment and disposal systems of Istanbul metropolitan area, *Water Science and Technology*, 44(2–3), pp. 31–38.

Eroglu, V., Sarikaya, H. Z., Ozturk, I., Yuksel, E. & Soyer, E. (2002) Water management in Istanbul Metropolitan Area. Paper presented at the *Third International Forum, Integrated Water Management*, Athens, Greece.

General Directorate of State Hydraulic Works (DSI) (2000) *Greater Istanbul Water Supply Project Stage II, Melan System* (Ankara: DSI).

Istanbul Water and Sewerage Administration (ISKI) (1994) *Istanbul Water Supply, Sewage and Drainage Master Plan Feasibility Report.* Istanbul Master Plan Consortium IMC, 6 volumes (Istanbul: ISKI).

Istanbul Water and Sewerage Administration (ISKI) *Annual Reports* (Istanbul: ISKI).

Istanbul Water and Sewerage Administration (ISKI) (2004) *The Adventure of Water in Istanbul.* ISKI Publication, No. 39 (Istanbul: ISKI).

A Regulatory Approach to the Jakarta Water Supply Concession Contracts

ACHMAD LANTI

ɟ

Background

Country Background

The Republic of Indonesia is the largest archipelago in the world. It has over 17 500 islands of which 6000 are inhabited (see Figure 1, the map of Indonesia). It has a total area

□   17,500 Islands
□   Population 213.6 millions
□   Largest archipelago country
    (2/3 of country area is seas)

Figure 1. Map of the Indonesia area

of approximately 1.9 million km^2 and in 2002 the estimated population was approximately 213.6 million, making it the fourth most populated country in the world. Its urban population of 74 million (35% of the total population) has a growth rate more than twice the overall country population growth rate of 1.5% per annum. Administratively, the country is divided into 30 provinces, about 400 districts (or regencies), 55 municipalities, 35 administrative cities and 16 administrative municipalities. In addition, it has two special regions and one special capital city district. The districts have become the chief administrative units for the provision of most government services. In 2001, life expectancy in Indonesia was 66.3 years, the infant mortality rate was 33% per 1000 live births (compared with 41.6 in 1997 and 35 in 2000), and the child mortality rate (for children under 5 years) was 45 per 1000 children compared with 48 in 2000. Its illiteracy rate is approximately 12%. The Republic carries a national philosophy 'Unity in Diversity' with five principles of state, i.e. in God we trust, humanitarian, unity of nation, people representation and social justice.

The country is endowed with rich natural resources: petroleum, natural gas, coal, timber and a number of minerals such as bauxite, tin, copper, gold and silver. These resources propelled Indonesia to a strong economic growth for some 30 years before it was hit by the Asian financial crisis in 1997/98 and by political turmoil in 1998. Its per capita GNP, which peaked at US$1120 in 1997, dropped sharply to US$580 in 1999, and recovered to US$690 in 2001. Indonesia is now emerging from its political and economic crisis and has undergone tremendous changes in its structural and political reforms since 2001. These include a major political liberalization, the 'Big Bang' decentralization on 1 January 2001, and an economic recovery and poverty reduction program. These have resulted in an upturn in economic growth, with a GDP growth averaging 4% over 2000 and 2001.

The city of Jakarta was founded more than 470 years ago, while the Republic of Indonesia's proclamation took place on 17 August 1945; Jakarta was also enacted as the Republic's Capital city. Jakarta is the capital of the Republic and is located on Java, the most densely populated island. It covers an area of 662 km^2 and in 2005 its population was approximately 9.9 million.

Table 1. Population of Jakarta, 1990–2019

Year	Population	
	No.	Annual Growth Rate (%)
1990	8223	
1995	8800	1.40
2000	9400	1.33
2005	9900	1.04
2010	10 400	0.99
2015	10 900	0.94
2019	11 200	0.68

Average population density in 2003 was approximately 14 700 people/km^2. Average household size in 2003 was approximately 6.7 people/household.
Source: The Study on the Revise Jakarta Water Supply Development Project, Final Report Vol. 2 Main Report Nihon Suido Consultants Co., Ltd., May 1997.

Overall trends in population and households, together with resultant growth rates, average population density and average households size are shown in Table 1.

Perusahaan Daerah Air Minum DKI Jakarta (PAM JAYA) is the water supply enterprise owned by the provincial government of Jakarta and it has the main sole responsibility for the provision of a water supply to the people of Jakarta metropolitan city, whilst the sanitation or sewerage system is under the separate management of the enterprise PAL JAYA. PAM JAYA operated the water supply system from 1922 until early 1998. Since 1998, the Jakarta Water Supply Service has been divided into two areas, Western and Eastern Sectors respectively. The water supply system in these two service areas is operated by two private operators. Ondeo, through the locally established enterprise, PALYJA, operates the Western Sector and RWE Thames operates, through the same set-up known as TPJ, the Eastern Sector, under 25-year concession contracts. The Ciliwung river, which flows south to north through the centre of Jakarta, is the boundary between the East and West Sector concession areas. In 2002, the water supply system in the West Sector produced 437 789 m^3 per day and received an additional 199 180 m^3 per day from the Tangerang Water Supply and another 11 664 m^3 per day from TPJ through a cross-boundary connection and the treated water was distributed to 312 879 customers by means of a distribution pipeline approximately 281 km long. During the same period, the East Sector produced 694 051 m^3 per day and the water was distributed to 336 550 customers by means of distribution pipelines 310 km long.

Past Condition of Water Supply and Expectation in the Future

In 1996 the coverage of service was at only 41% while the use of non-revenue water was high and no less than 57%. The people who did not have a connection were using groundwater as a water source. Groundwater was not only used for domestic consumption but for other uses as well, such as for commercial and industrial purposes. As a result, over abstraction of the groundwater occurred, which has caused its depletion. If this continues unabated the groundwater depletion will cause deterioration of the environment, for example, causing land subsidence, which has been occurring at a rate of 2.8 cm per year. The government aims to improve the water service for the city in order to achieve potable

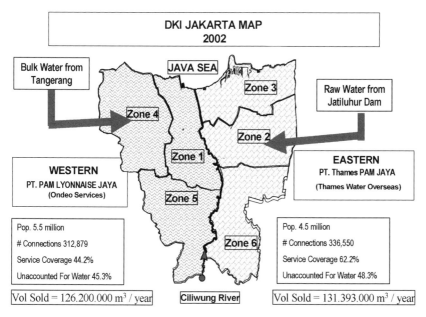

Figure 2. Map and statistics about the DKI Jakarta area. *Source:* Jakarta Water Supply Regulatory Body.

water quality and at the same time prevent even more deterioration of the environment by reducing the abstraction of groundwater and promoting the use of the municipal water supply by the population.

Prior to negotiating with the two private operators, the city government aimed to achieve a considerable improvement in the water supply service. In 2002, through a private sector participation (PSP), some improvements have been made, as illustrated in Figure 2. Groundwater abstraction was to be reduced to prevent its depletion and the occurrence of land subsidence. Service coverage was also expected to rise from 41% in 1996 to 70% in 2002. NRW was to reduce from 57% down to 35% in 2002. As a result, the volume of water sold would have increased nearly twofold during the same period (Figure 3).

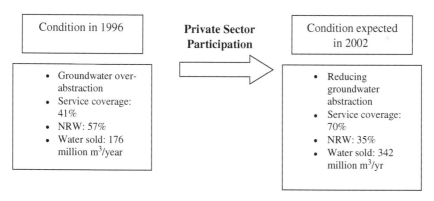

Figure 3. At the start of negotiations, conditions were expected to have been better after 5 years. *Source:* PAM JAYA Negotiation Team Presentation (1996).

Expectation in 2002	Achievement in 2002
• Reducing groundwater abstraction • Service coverage: 70% • NRW: 35% • Water sold: 342 million m³/year	• No significant reduction in groundwater abstraction • Service coverage: • West 44.17% • East 62.17% • NRW: 44% • Water sold: million m³/yr • West 126.20 • East 128.95 • Total 255.15

Figure 4. Comparison between expectation and achievement of the PSP after 5 years. *Source:* PSES Report of Palyja and TPJ.

In spite of the presence of two international private operators, the realized access to water supply does not appear to be in accordance with what was planned before PSP. The achievements of the private operators after five years is shown in Figure 4.

There were a number of reasons for the unconformity. First, was the occurrence of the Asian monetary crisis, resulting in a huge dive in the value of the local currency from Rp.2200 to between Rp.12 000 to Rp.8500 per US$1, before and after five years of the PSP respectively. Second, due to affordability for consumers during the crisis, the government introduced a policy of not allowing PAM JAYA to adjust the tariff for the first three years of the PSP, meanwhile the prevailing inflation during the period reached approximately 120%. However, in order to catch up with inflation, the tariffs were adjusted three times, i.e. 1 April 2001 by 35%, 1 April 2003 another 40% and 1 January 2004 by 30% respectively.

Third was the ambiguity status of PAM JAYA employees seconded to the operators, which represented approximately 90% of the total work force and to some extent this has been influencing the performance of the companies.

Finally, since these two concession contracts were the first types of their kind in Indonesia under the PSP scheme, each contracting party did not completely fulfil their obligations and rights, and sometimes it appeared there was a lack of 'good faith and best endeavour'.

Meanwhile, the average water consumption is fairly stable among the domestic users. Per capita consumption in 1995 was 156 litres per day, and in 2001 it was also approximately 156 litres per day.

The figures in Table 2 show a comparison between a number of key indicators of water supply provision, before PSP and after PSP until the end of the respective concession contracts.

One of the key policies of the government is to provide clean water for the poor through public taps and it is concurrently introducing a cross-subsidy policy by easing water tariffs

Table 2. Comparison between targets and realizations until end of concession period

	Total concession		Service coverage		NRW		Water sold (millions)	
	Target	Realization	Target (%)	Realization (%)	Target (%)	Realization (%)	Target	Realization
1993		324.433		38		53		158
1994		349.849		38		52		168
1995		362.618		39		57		165
1996		393.746		41		57		176
1997		428.764		42		5		191
1998	470.674	487.978	49	43	50	58	210	181
1999	571.776	541.630	57	43	47	54	244	208
2000	653.885	562.255	63	48	42	48	281	228
2001	597.174	610.806	50	51	47	49	236	237
2002	636.461	649.429	53	52	45	47	250	255
2003	675.534	690.456	54	56	43	45	258	274
2007	796.738		75		31		297	
2012	847.774		89		26		322	
2017	864.511		100		26		337	
2022	879.511		100		26		353	

Source: Jakarta Water Supply Regulatory Body, 2004.

for low-income consumers. In addition, government provides subsidies by extending pipe distribution to areas where the urban poor live. There are several types of small-scale water service providers in Jakarta to serve the urban poor:

- Public hydrant: the water source is 'treated water' from the utility and conveyed to consumers by means of push carts.
- Water tank and public hydrant: water is transported to the public hydrant by means of a tank car and subsequently distributed to consumers by push carts. The service area is located outside the utility network.
- Water resellers, where the households that have connections resell the water to those who are not connected.
- Separate network with a groundwater source and deep well: households are connected to the distribution pipe system, separate from the utility network.

Wastewater and Sanitation

According to the Joint Monitoring Program on access to water supply and sanitation, 76% of the urban population in Indonesia has access to basic sanitation. SUSENAS data indicate that of these, 68% has access to private basic sanitation facilities; 14% use shared facilities, 8% use public facilities, and 10% use other non-specified facilities.

Septic tanks are by far the technology of choice in urban areas (59%). In Jakarta alone, there are approximately 1 million septic tanks. Other means for the final disposal of household human wastes include pits (21%), rivers and lakes (13%), and ponds, with rice fields and others taking up the remainder.

The coverage of sewerage service in Jakarta (by PAL JAYA, the Jakarta Sewerage Enterprise) was only at 1.90% in 2000, mainly serving high-rise buildings and a small number of households. The sewage is treated by means of an aerated lagoon or facultative lagoon that has a capacity to treat $11\,700\,m^3$ domestic wastes daily. Some apartment buildings treat their waste in a communal waste treatment system.

In general, there has been no significant improvement in domestic waste disposal in Jakarta: septic tanks 39%, pit latrines 20%, and others (not indicated, possibly discharged directly into the rivers) 59%. In areas, where the Kampung Improvement Project has been implemented by constructing communal waste treatments, there is approximately 15% improvement with regard to domestic waste disposal.

With regard to the national policy on WSS (water supply and sewerage system), a law on water resources development was recently approved by parliament, after going through very lengthy and tedious discussions for about a year. The new Law No. 7 of 2004 on Water Resources has been passed. It deals with the provision of sustainable water resources, management of water supply and wastewater, and participation of the private sector, as well as cooperatives. A supplementing government regulation on water supply (Regulation No. 16/2005) and Presidential Decree No. 67/2005 on Development of Water Supply System, were issued to formulate the role, responsibility, rules, and procedures on how the Local Government should manage the water supply and wastewater system and how to introduce the PSP in water supply development including the establishment of Water Supply Development Supporting Agency/Badan Pendukung Pengembangan Sistem Penyediaan Air Minum (BPP-SPAM). Under the clause 40 of the Regulation 16/2005, it is noted that WSS should be managed as one integrated system in order to protect raw water for domestic, industrial, and commercial uses. In a number of big cities such as Bandung, Jogyakarta and Medan, the WSS is being managed under one local enterprise. However in Jakarta, due to historical reasons, it is under separate management (i.e. PAM JAYA and PAL JAYA respectively). This paper is not going to discuss the pros and cons of the above two systems. Hence, the Jakarta Water Supply Regulatory Body (JWSRB) is only involved in the water supply system.

The Process and Implementation of Private Sector Participation

Negotiation of Contracts

In June 1995, the Minister of Public Works (currently the Ministry of Settlement and Regional Infrastructure) received initial guidance from the President of the Republic of Indonesia, that Jakarta needed privatization of the water supply to enhance the provision of a clean water service to the people of Jakarta. At that time, the local government operator, namely the Jakarta Water Supply Company (PAM JAYA) was facing a very difficult situation due to lack of an investment budget to increase the coverage ratio which was low at only 41%. Due to the fact they had no experience of such large-scale privatization in the past, and that there was no applicable law on it at that time, the Ministry, after consultation with local government of Jakarta, chose two reputable international operators, Thames Water International in the UK and Lyonnaise des Eaux in France. However, it was necessary to choose other private operators in case these two companies failed to meet the Indonesian government criteria concerning the aim of improving water supply services. Based on the above criteria, the Ministry issued two letters of intent (LOI) in June 1995 and August 1995 to KATI (the joint operation of Thames Water International and a local company) and to GDS (the joint operation of Lyonnaise des Eaux and a local company) respectively. They asked the companies to execute feasibility studies (FS) concerning the possibility of Private Sector Participation (PSP) to develop clean and potable water services in Jakarta by dividing the city into two parts, Eastern and Western, and granted 25-year concessions for each. FS reports were submitted by GDS and KATI in March 1996 and May 1996 respectively. Upon the recommendations of the Coordinating and Evaluation Committee, the Minister approved the FS reports as the baseline towards further negotiation to conclude the Cooperation Agreements. Finally, in June 1997, Cooperation Agreements were signed between PAM JAYA and GDS for the Western part and KATI for the Eastern part. After compliance with a number of preconditions, the agreements became effective on 1 February 1998. The two concessionaires with headquarters in France (ONDEO Services) and in Germany (RWE) are now known as PALYJA and TPJ respectively.

As far as the selection method is concerned, the use of an unsolicited process was granted due to the non-existence of laws and regulations, in addition to them being considered as pilot projects of PSP in the water supply and sanitation sectors. During negotiation, activities were underway; the instruction of the Minister of Home Affairs (MOHA) No. 21/1996 was issued. It has become apparent that all activities of the pertinent 'unsolicited process' of the Jakarta concession contracts are in line with the said MOHA instruction.

Implementation of the Contracts

Establishment of Jakarta Water Supply Regulatory Body. The first three years of the contracts were very tough due to the Asian Economic Crisis and the change in the government regime, in addition to the unwanted social unrest that occurred in Jakarta. The above contracts were finally reformed on 22 October 2001 by signing new 're-stated cooperation agreements' (RCA). One important aspect of the RCA was establishment of the JWSRB through Governor Decree no 95/2001 in September 2001. The study and design of the respective JWSRB was originally initiated by using the IBRD assistance to the Jakarta local government that was prepared by NERA of Australia in June 1999. After the first three years, which ended the first term of the JWSRB, a new regulation for the second term (2005–2008) was issued under the Governor Regulation No. 54/2005 specifying the enhancement of the role and the authority of the JWSRB. Unlike the first term, the chairman and members of the Board were chosen through the public recruitment system.

Rate re-basing. Due to the persistent unsolved agreement of both parties on mounting MOF debt and the operators' shortfall, more recently in November 2003, the Ministry of Settlement and Regional Infrastructure (now, Ministry of Public Works), together with the Jakarta government, set up an Independent Combined Expert (ICE) team, which includes the Regulator, to execute a 'rate re-basing' exercise and establish a new basis for future water tariffs. The ICE team conducted the rate rebasing exercise and, in mid February 2004, presented the results of its exercises. However, no agreements could be reached. The Jakarta government then consulted with the Ministry of Public Works, and both the governments formed a joint team to facilitate both parties to arrive at an agreement on rebased water charges. Again, this attempt ended in failure. As a result, at the end of September 2004, the DKI Jakarta government let PAM JAYA and PALYJA/TPJ go back to the negotiation table to resolve the issues concerned with rate rebasing, but again, to no avail. Finally, both parties came to JWSRB and requested mediation. In December 2004, PAM JAYA and PALYJA reached an agreement for a new rebased water charge. However, TPJ failed to reach such an agreed rebased water charge, though it was one of the prerequisites for the implementation of an automatic tariff adjustment. After prolonging unsolved disputes, by the end of November 2005, both parties finally accepted the opinion of JWSRB on the rate rebasing effective from 1 July 2004. This was accomplished during the plenary meeting before the High Executive Meeting of the Jakarta Provincial Government chaired by the Governor, before the first implementation of an automatic tariff adjustment in January 2005.

How to strike a balance between public and private interests. A key aspect of the approach of the JWSRB in the complicated and rather difficult situation as described above, is how to strike a 'balance' between public and private interests, thus enabling the cooperation needed if reliable technical and service standards at affordable tariffs are to be reached for Jakarta's 10 million inhabitants. This balance is required if the different objectives of the key players are to be reconciled in a sustainable way. This paper describes the challenges faced in achieving this, since the establishment of JWSRB in September 2001.

When a government itself is responsible for all aspects of infrastructure, achieving a balance between the various players might be regarded as less vital, the assumption being that the Minister will act in the best interests of all. Governments can, as sole provider, determine what is needed, although this is increasingly open to debate as the voice of the customer is seen as increasingly important. In the search to find more effective institutional set-ups for infrastructure provision, there is increasing emphasis on separating roles so that better incentives can be designed for each party and their performance better scrutinized. This is the driving force behind PSP.

The main balance to be struck is between the profit-motivated private sector and the public interest, i.e. an affordable service. If this is not achieved, the service will not be sustainable, the private sector may withdraw, on-going disputes will occur and the promise of a greatly improved service that PSP offers may not be realized. The remainder of this paper looks at these issues as they have emerged in Jakarta over the last four years.

What Interests Need Balancing?

Is the Public Interest More than What Is Written in the Cooperation Agreements?

An initial reply to this question is that it is important to understand why a 'Regulator' has been introduced in Jakarta. The classic answer is that it is in order to ensure that services are provided at a fair price and that value for money is obtained. At the most basic level, the public is interested in receiving a service at the lowest possible price. However, it is worth emphasizing that most recent costumer satisfaction survey shows that the public is increasingly demanding that they receive a reliable quality of services from the operators. In so far as generalizations hold true, the public wants a reliable, if not continuous supply of water. They want easier payment facilities, they do not want to be asked to pay for water which they do not consume and they do not want to be asked to pay for water when their neighbours are receiving it free from illegal connections. In addition, many say they want a supply that can be drunk directly from the tap. Translating these desires into concrete actions that the public perceive are in their interest is a major challenge for the JWSRB. Identifying and accommodating changes in the public interest will be even more challenging.

Despite the above points, in Jakarta the main reason given in official circles for establishing the regulatory body was the obvious need for an independent, impartial 'regulator' between the private operators and local government, in this case represented by PAM JAYA.

Clause 50 of the original Cooperation Agreements of 6 June 1997 and strengthened in the RCA signed on 19 September 2001 confirms that such an institution will be established, given both that government consists of a number of stakeholders/parties and the private sector concern that government would be unable to rule fairly on disputes involving itself. It was not surprising that the establishment of an independent regulator was a very high priority for the concessionaires. The presence of a credible independent dispute resolution mechanism also remains a prime concern for private operators.

However, there were a number of other important reasons for establishing the regulation of water concessions in Jakarta. These reasons were not widely acknowledged in local government because they were primarily matters serving public interest, hence they did not receive the same promotion by the private sector as the need for an independent regulator. Nevertheless, the importance of these other reasons will increase over time. They include:

- Countering monopolistic tendencies in the pricing-service mix of the operators (including the private operator's public partner, PAM JAYA), and include guarding against the possibility that the operators will make windfall profits or enter arrangements to further strengthen their monopoly position.

- Providing a mechanism for regulating 'externalities', for example, to control groundwater abstraction, catchments and water resource protection or waste discharges, the costs of which may not otherwise be captured in the water tariff.
- Making information on the market for water more accessible, accurate and affordable, thereby protecting consumers and the public against information inadequacies and the possible abuses that may result from it.
- Ensuring that the water services to all communities is a matter of social policy, so that it is not just the rich and industry/commerce who benefit from the service.
- Creating an independent advocacy for the rational pricing of water, without which the service is to be captured by few, while the entire community pays in some way, generally through taxation.
- Countering the unequal bargaining power that PAM JAYA, local government and especially the community may suffer at the hands of a determined, well-funded and experienced concessionaires; in effect to lower the transaction costs.
- Ensuring the long-term planning and coordination among concessionaires (if any) as well as between concessionaires and other public services, so that the service delivery can be cost-effective and more efficient.

While these reasons provide the technical or economic rationale for the creation of a regulator, it is worth noting some of the less tangible explanations. Besides 'the public interest' rationales above, it might be argued that there is also 'private interest' at work.

Whatever explanation is given for regulation, worldwide experience shows that an increasing number of governments are establishing specialized regulatory agencies. The proof of this can be seen just by examining the agencies listed in the World Bank's *International Directory of Utility Regulatory Institutions*. Since just a few listings in the first edition of 1996, the 2000 edition now contains over 900 agencies in 185 countries.

Is the Private Interest More than just Profit?

In listing what the private sector is interested in, it is tempting to just stop at 'profit'. There is also a tendency for many people, perhaps because of the experiences of recent times in Indonesia, to assume that profit is a bad thing. First, in relation to profit, it is important to understand that profit in itself is not a bad thing, as some opponents of PSP would like people to believe. The market system has become the dominant economic system worldwide because of the incentive to improved performance that profit gives. Human advancement in many fields has been driven by profit. Its use in a system to improve the provision of infrastructure is only natural. It might be said that it is because of profit that we have most of the advances of the modern world.

Second, what is important about profit is its amount in relation to the amount of own funds put at risk by the investor and the return received in relation to the risks faced. In Jakarta a return of 22% (measured as an Internal Rate of Return, or IRR, on equity over the life of the concessions) has been agreed for the risks currently faced. Profit is of more concern to the regulator when it becomes excessive, or when there is not enough of it to ensure that the operator will continue to invest in the business.

Third, besides matters of profit, there are a number of other aspects of interest to the private sector. Among these might be listed: continuity of effort, employee satisfaction, maintaining a good reputation, accomplishing a mission and sustaining the vision of the way the mission will be accomplished.

Hence, to achieve a satisfactory balance from the perspective of the private sector, the regulator needs to be aware of the amount of profit needed by the regulated firm, the risks faced by the firm and the return that the investor might receive for investment with similar risks. While profit is at the end of a prime interest of the private sector in participating in infrastructure projects, the way in which a profit is made should be a concern.

Good regulation can help to ensure that the private sector's interests are balanced fairly with those of the public interest. In the long term, good regulation is likely to show that what is good for one is also good for the other. But how can this balance be achieved? As an introduction, the next sections consider who regulates what and the parameters for judging whether the regulatory system will work satisfactorily.

Who Should Regulate What?

As mentioned earlier, the Jakarta concession contracts contain agreements to create an independent regulator. At this stage their role is mainly to mediate on disputes between the parties. A frequently asked question from the local government of Jakarta has been 'why do you need to create a new institution and why can't another branch of local government or PAM JAYA themselves, to do it?' The very term 'regulator' has been challenged, quite rightly with the observation that the government has always regulated, so why is a new special body needed?

It is correct that governments do regulate. Examples can be found in regulations by many forms of governmental institutions. These range from local governments (an example being building permits or environmental health), self-regulation (the professions and the Jakarta Stock Exchange), parliament (fuel subsidies), central government (currently the energy and telecommunications sectors), or through special agencies (Bank Indonesia in the banking sector, or the anti-trust Commission-KPPU). The question is more 'what form of organization is the most effective regulator' rather than 'Should there be regulation?'

The recent experience in Jakarta has highlighted the difficulties of converting an existing government operator, PAM JAYA, into a regulator. As a matter of principle it is suggested this be avoided in the future. The second reason for not using the government-owned partner is the more obvious one of the conflict of interest that the partner faced from being a party on the one hand and a regulator on the other. These considerations suggest the need for creation of a specialized agency. In fact, this has been the worldwide response. Independent agencies have an advantage over government as regulators for a number of reasons. They offer continuity of policy across changes of Minister and government, they are able to better combine the functions of adjudication, policy making and enforcement as well as that they can develop higher degrees of expertise and they are, and can be seen to be, independent.

The next question concerns the roles and responsibilities of the regulatory body. Should it be entirely responsible for (1) policy making; (2) elaborating detailed standards and targets; (3) monitoring compliance with norms; (4) tariff setting; (5) adjudication/mediation; and (6) enforcement/imposition of sanctions, or should the responsibilities be divided between local government, PAM JAYA and the regulatory body?

The fact that regulation has grown out of the cooperation agreements has largely determined the current agreement, which is that the regulatory body will mainly play a mediator role and focus on technical issues. As an example, the policy decision relating to tariff changes is therefore still retained by local government because its determination is based on current political criteria, although in time this should become a matter for regulatory decision based on the clearly defined criteria of Government Regulation (PP) No. 2/1998. The allocation of minimal responsibilities is considered best at present until the institution can gain the confidence of government by building its expertise and proving its ability to perform.

Another concern regarding responsibilities is the scope in relation to all other areas that firms suffer from regulations, for example, health and safety, environment, anti-trust/pricing etc. While the details are yet to be worked out, the obvious policy is to confine the regulatory body to those of 'economic regulation'. This means that activities will be focused on ensuring that technical targets and service standards are appropriate for the tariff charges. The role in relation to other areas will be much more that of 'coordination'.

What then are the key areas on which performance of the regulatory body might be judged? The following section discusses these, all of which require achieving a balance between numerous forces often at odds with each other.

How to Achieve a Balance: The Five Parameters

During the last four years, the JWSRB has undergone a 'learning-by-doing' process, although there are a number of parameters that are just beginning to develop for judging whether the system is effective. An important step was taken in October 2002 at a Seminar on 'Regulatory priorities in emerging markets' held at the Grand Hyatt Hotel in Jakarta at which the principal speaker was Sir Ian Byatt, formerly the Water Regulator for England and Wales 1989–2000. The Head of JWSRB analysed the performance of his new Office by reference to the following parameters:

- *Mandate:* Does the JWSRB have a clear mandate to perform all its tasks and functions; in other words, does it have sufficient legislative authority?
- *Accountability and independence:* Is it accountable to key stakeholders; in other words, is there an appropriate system of accountability? At what level of degree the control of local government and local parliament over the JWSRB for ensuring its independence?
- *Transparency:* Is its operation transparent, is information readily available and are procedures fair, accessible and open? Does it collect the right information on costs and performance?
- *Expertise and credibility:* Does the JWSRB have or does it act with sufficient expertise to have shown its credibility in attracting investment whilst protecting consumers and PAM JAYA interests?
- *Efficient and fairness:* Is the system efficient at delivering its objectives? Does the JWSRB have shown a reputation of fairness both to concessionaires and consumers? Does it have a clear process for decision taking?

The Legislative Mandate

This criterion is fundamental for ensuring that the regulator cannot only go about his task with the full backing of the state, but also that the position has the support of the public. The position of the regulator needs to be firmly entrenched in the legal and administrative system for it to withstand the many challenges it will face. It would be best that its authority stems from the body most closely representing the community, that is, the democratically elected parliament (the local parliament, DPRD or even better, the national parliament, DPR).

The most recent Governor Regulation no. 54/2005 is considered as a temporary measure until the position of the regulator can be strengthened through drafting, discussions and issue of a PERDA (a local regulation). In fact, in discussion with the provincial government on the positioning of the JWSRB, it seems clear that national legislation is needed to give full legitimacy to such bodies and to ensure some uniformity across the country. However, in spite of the recently enacted Law on Water Resources (no. 7/2004) and its Government Regulation (no. 16/2005), the expected establishment of an overarching National Regulatory Body (such as JWSRB), has not as yet properly materialized.

One key point those drafting and approving the legislation must ensure is the setting of clear objectives. Without clear objectives it will be difficult for the regulator to claim or prove that the mandate has been fulfilled. Given the sensitive nature of infrastructure services, the generally poor understanding among legislators of key issues and the propensity for legislators to intervene at the micro level, there is an obvious temptation to frame the legislation in general terms.

The quality of the legislation in ensuring the independence and accountability of the regulator is the other important area. This quality will be borne of careful drafting and wide consultation on the detailed provisions, some of which are considered below.

Accountability and Independence

Just as unchecked monopolists are likely to favour their own interests over others without a system of checks and balances, the regulatory agency needs to be held accountable. Again, it is a matter of balancing, in this case that of independence with accountability. Control does need to be exercised over the regulator, although that control is not of the type traditionally reflected in the 'command-line' approach to the exercise of government power.

Key issues in discussions concerning the JWSRB relate to the degree of control that the governor and legislators have over the regulator, for ensuring independence and the mechanism by which the body will be held accountable. Experience shows that neither can be discussed without the other. In promoting the need for independence of the body, accountability must be talked about. It is also useful to point out that the presence of an independent regulator can relieve the executive and legislators of some of the pressures in relation to unpopular decisions (such as the issue on tariff changes from time to time).

Based upon the previous four years of experience, the suggested arrangements for the JWSRB, although not yet fully accommodated by the current Governor's Regulation of 54/2005, April 2005, include:

- Supervisory control (as opposed to intervention) exercised through the relatively detailed prescription of tasks and responsibilities in the Decree. However, this task, is not as yet fully controlled by JWSRB, because overlapping responsibilities with those of PAM JAYA are still in existence.
- A requirement for written procedures to be prepared by the regulator and approved by the authorities in due course, enabling the system to perform effectively. As yet the system is not fully implemented, except for the procedures attached to the tariff changes.

- The appointment of regulators by the executive with legislators' approval based on selective, published and publicly acceptable procedures, towards achieving better representation and accountability. The current appointees were basically selected by the local government, by using criteria on educational background and related pertinent experiences.
- Distinctions being drawn between areas in which the regulator is authorized to make 'decisions' as opposed to those in which the regulator would make 'proposals' (the decision being a matter for executive or legislators). This is an arduous matter, as had been proven at least three times that whenever the JWSRB issued decisions on a number of important aspects, they were always challenged, notably by the private operators without any acceptable reasons. Furthermore, there were no follow-ups whatsoever related to the procedures as set out in the contracts, that is final decisions to be drawn up through what is known as an 'Expert Team' to be established, by the parties as an 'ad hoc arbiter'.
- Annual reporting to the executive/legislators and the public audit of regulator financial and operational performance which will enable enhancement of a greater 'sense of accountability'. Apparently, these had been no problem to execute.
- Removal of regulators is to be enforced only on prescribed grounds with legislative approval. The procedure for removal with prior legislative consent is clearly described in the said Governor decree.
- Public consultation and disclosure concerning decisions and proposals are to be mandated to the JWSRB. Under the patronage of JWSRB, one independent consumer forum at the provincial level (FKPM) and five subordinated ones at mayor levels (KPAM) were established. These forums are quite effective in capturing the consumers' complaints as a baseline for further remedial actions to be carried out by the operators.

As previously mentioned, the current JWSRB establishment has an interim status for three years, and it is hoped the above arrangements can be fully accommodated thereafter. With regard to whom the regulator is accountable, it appears that legitimacy of the regulator is unlikely to be strong unless it is directly accountable to a body which itself is representative of the community, that is, to the DPRD. This is another reason for upgrading the legal instrument establishing the JWSRB to that of PERDA status and for moving away from the contract-based regulation currently foreseen by some parties.

Transparency and Due Process

Acting in the public interest requires that particular attention must be given to informing the public of what is being done on their behalf. Transparency is thus essential if the regulator is to be seen to be legitimate and truly acting in the public interest.

Transparency will lead to procedures that are seen to be fair, accessible and open. Key stakeholders cannot then complain that there is bias in how they are regulated. A fair deal becomes a reality for both investors and operators while the community has the opportunity of participating in decision-making.

Transparency is also called for to ensure that consultations between the stakeholders are productive and, given the noted difficulty in determining the true interests of the stakeholders, an environment is created in which the interests can be exposed.

Some practical steps to improve transparency are being discussed for the Jakarta arrangements. During the four years since its inception, the JWSRB has undertaken the following:

- Under its patronage, in January 2002 an independent forum of consumers, together with all stakeholders concerned at the provincial level (known as FKPM), was established. A year later, five more forums of grass roots consumers' committees on water supply at mayor levels (KPAM) were also formed; in December 2003 the Water Voice system was launched, giving a new impetus to this dimension.
- In early 2005, the JWSRB created a public information centre through an Internet website (www.jakartawater.org), where the public can access information concerning the cooperation agreements, the performance of the concessionaires and the consumers voice, as well as the policies and procedures of the regulatory body.
- To a limited extent, the JWSRB is advising the public through electronic media and newspapers about the performance of operators as well as information on tariff adjustments from time to time.
- Annual budgets are to be presented and approved by the Governor and year end audits.
- The JWSRB has submitted Annual Reports four times, including an assessment of both operator and regulator performance to the Governor before it was made available to the public.

There are practical limits to transparency and participation in decision-making. A balance must be struck between allowing key stakeholders to participate, and the regulator complying with his responsibilities to ensure the successful on-going operation and development of the water supply service.

Expertise and Credibility

There are likely to be many occasions when rapid decisions or proposals must be made on incomplete or changing information. These situations call for expertise. The application of expertise by the regulatory body also enhances its credibility and power, and to some extent can be used to 'push through' decisions without time-consuming explanations and analysis, which may be of doubtful benefit anyway given the range of variables.

Although the availability of expertise helps to establish the credibility of the regulator, the application of expertise will also enable better explanations to be given and more rational work processes established.

The challenge for Jakarta and indeed all the new regulatory bodies under consideration is in acquiring the expertise. In this regard a number of strategies can be deployed. The tendency to create large new organizations staffed with ex-PDAM or civil servants needs to be resisted This does not mean that all these people do not conform to the required expertise, but rather, that the best expertise should be recruited, either from the public or private sector. It is also based on the observation that ex-operators do not necessarily have the knowledge, skills or attitude to become good impartial regulators. A balance between use of in-house expertise and consultant expertise is now being implemented.

Efficiency in Achieving Objectives and Fairness

This efficiency can be measured in terms of how the regulation is achieved and in the results that the operators are able to deliver. Clearly, the credibility of the regulator will be enhanced if its operation is perceived to be more efficient.

The method of regulation may, for example, involve the use of large numbers of staff collecting and processing large amounts of information under a very 'command and control' regime. Alternatively it is now being implemented by self-regulation. The difficulties seen here especially relate to the 'what if' case, and to the objectives. Because it is very difficult to know what would be the outcome if a different method were used, it is rather difficult to claim that the regulation used is efficient. That also first assumes that the regulator's objectives are clear and measurable. As noted previously, the mixed bag of objectives handed to the regulator by the legislators may make this measurement impossible.

Evidence of efficiency might also be sought in the results the operators produce, in effect the degree to which the technical targets and service standards laid out in the Schedules to the Cooperation Agreements are met each year. Again, the problem here lies in what alternative could have been produced, and in knowing whether the targets and standards provide the optimum result for the community anyway.

Benchmarking against similar operations, both nationally and internationally will provide one key to demonstrating efficiency and simulating market forces for the operators). The successful establishment of the benchmarking system by PERPAMSI (Indonesia's Water Supply Enterprises Association) is therefore very important, as is the preservation of the separate operations in the east and west of Jakarta. International networking among regulators is also expected to provide relevant information. Although the operators may object that every operation has its peculiarities and therefore should not be compared, there appear to be few other avenues open to regulators to impose the discipline that the market would otherwise provide.

To ensure fairness to both operators and consumers, the following methodology and mechanisms are being carried out:

1. The operators' interests need to be fairly assessed in an independent manner by the regulator. This endeavour is not as yet properly implemented due to the existing overlapped/co-existent functions between the JWSRB and PAM JAYA.
2. Regular meetings are also being conducted with consumer input (FKPM and KPAM), regarding their complaints/dissatisfactions of services being delivered and they are demanding the operators remedy them in due course.
3. The JWSRB in close collaboration with the independent surveyor has completed the consumer satisfaction surveys for the years 2003, 2004 and 2005 in order to capture:
    (a) The level of existing services;
    (b) Types and natures of complaints;
    (c) Consumers expectations for service improvement.

Responding to the Challenge: Learning to Regulate by Doing

The response to these difficulties, exacerbated by the 'monetary crisis', was to largely 'freeze' the incentive mechanism concessions by the introduction of a 'Transition Period' for the whole of 2002. It was agreed that this period was necessary to gain better information about the system, the reality and reasonableness of expenses on OPEX and to rebuild trust between the parties.

In effect, the regulation of the operators in this period was similar to a 'management contract' regime, where incentives are minimal, as are the risks. Quarterly submissions of information were made, including proposed expenditure in forthcoming quarters. This involved negotiation between the parties (with mediation by the regulator) through the mechanism of the Performance Evaluation and Review Group (PERG) meetings conducted every month. They were useful in revealing the real and reasonable costs required to operate and develop a sound and productive system. To improve information exchanges and reliable inputs to PERG, five Counterpart Staff were placed by PAM JAYA (using recruitment criteria by Regulatory Body) in each of the concessionaires' offices. 'Open book accounting' was practised.

A balance had to be struck between involving PAM JAYA in management affairs and leaving the project manageable in the eyes of the private operators and their bankers. Unfortunately, the design of performance incentives has been difficult to realize in this transition period, although every three months the government will have the opportunity to ask for efficiency gains in the succeeding quarters if it sees fit. That may be the price to pay for a better understanding between the parties and for opportunities for the regulator to begin with accurate information at the end of the transition period.

By the end of the transition period (i.e. early January 2003), both parties should have agreed on new water charges, when all previous money owed from the past had been accrued, and the current benchmarks, the variable costs of OPEX as well as the forthcoming five-year investment period up to the year 2007, had been fixed. In fact this ambitious milestone was not reached, due to the lack of trust and to a certain extent a number of data that were not available, mainly due to the operators' classified documents. In order not to prolong this uncertainty, the Ministry of Settlement and Regional Infrastructure (now Ministry of Public Works), together with the Jakarta Provincial Government, took an initiative by requesting that the Asian Development Bank should deploy an Independent Combined Expert Team (ICE Team), consisting of a reputable international consulting firm working together with the local firm, to finalize the said rate re-basing exercises, and this was scheduled to be concluded by February 2004. The ICE Team submitted their final reports in early March, and now it has become apparent that the rate re-basing findings have not been accepted by all parties concerned, including JWSRB. In spite of TOR of ICE Team the final decision shall be worked out through the mechanism of involving the Governor as an arbiter. Certainly this system is not worthwhile since it is actually the task of the regulator to decide. In the Legal Adjustment of the Cooperation Agreements, the regulator insisted on having more authority as stated in Governor Regulation No. 54/2005 concerning the role and responsibility of the JWSRB, particularly in regard of decisions on adjustment of financial projections by taking into consideration affordability for consumers to pay water tariffs. In future, presumably under the amendment of the cooperation agreements, this arrangement must be replaced by the correct one, namely to give more authorities to the JWSRB.

Some Lessons Learnt by the JWSRB

The experience of the Jakarta concessions to date has highlighted a number of key issues that any regulator needs to be aware of and find ways around. These include lessons concerning the technical issues of regulation as well as some concerning the establishment and operation of the regulator. They all involve balancing a range of interests.

Technical Matters: Information, Benchmarking and Competition

After a year or more of re-negotiation of the cooperation agreements, the following points in relation to regulation stood out:

- The difficulty of determining what are fair expenses for the operators, especially when the initial prices were not set by market bids and information from peer operators is not readily available.
- The importance of accurate operational information and an understanding of how investment affects performance.
- The asymmetry in access to information, in favour of the operators, as it is they who operate the system day-to-day.
- The importance benchmarking of peer operations will play in simulating competition for the concessionaires.
- The difficulty of really simulating competition, since it is not as easy as it might seem. It is clear why many say that there is ultimately no substitute for the discipline of the market.
- Less competition in acquiring the services will need the presence of good regulation.
- The difficulty in ensuring that tariffs are sufficient to ensure there will be continuing investment in improving water supply systems.

Challenges in Establishment and Operation of the Regulator

The key challenges for the regulator during the previous two years and those envisioned in the future, are reflected in the areas that need attention:

- The regulator taking early actions to gain credibility and to develop public support.
- The regulator being seen to use resources efficiently.
- Minimizing the decisions the Regulatory Body must take immediately while building capacity.
- The provision of start up technical assistance.
- The prospect of achieving more sustainable cooperation for the future.

Putting the customer first. Tariff increases were granted on 1 April 2003 and 1 January 2004 respectively, in order to catch up with and start to provide investment funds for developing the Jakarta water systems. The JWSRB had performed a public campaign/socialization to gain public support to achieve these increases. This has been an indispensable task, bearing in mind that regulator is there to protect the public interest, therefore it is vital that the public sees this and so supports the regulator. This recognition also aids independence. The behaviour of the regulator will be vital in maintaining this support.

1. All decisions by the regulator should be arrived at through a transparent process. Major decisions affecting the public should involve consultation and public

hearings. The processes of the regulator should be established early on and the public informed of them.

2. Through the established voluntary community consultative groups (i.e. KPAM, FKPM, YLKI) there should be communicating forums as a part of 'water voice system' to ensure that a channel for communications is created.

3. Further public support will be gained if the regulator focuses early on matters that are of greatest concern to the public. Quality of water, levels of service, price and convenience immediately come to mind. This undertaking is to be implemented through regular meetings with consumers and other related stakeholders, and more importantly, subsequent actions are urgently needed to be delivered by the operators in due time. The actions and their results should be publicized.

4. The role of the regulator would also be advantageous. Higher levels of government will need to pass laws enabling the regulator to fulfil his role. Upgrading the SK to PERDA will be important. In the process local government should agree to be bound more by the decisions of regulator, and especially in relation to tariff.

5. The regulator must remain independent. Any impressions that the position is not at arm's length from the executive and legislative branches must be avoided. The impression that the regulator has been 'captured' by the operators must also be avoided.

Use resources effectively. The expertise and resources available to the regulator will be relatively limited for some time. However, its tasks must be addressed with quality inputs if credibility is to be built. It has been the policy of JWSRB that the key points in achieving effective use of resources are likely to be: (1) outsourcing to consultants; (2) avoiding complicated disputes; and (3) putting the operators to work.

Becoming embroiled in disputes between the different parties will consume an enormous amount of resources and distract from other important matters that must be attended to. In the transition period the JWSRB has chosen 'a low profile' attitude, enabling PAM JAYA and the concessionaires to resolve their differences without detailed intervention by the regulator.

Building the capacity. Clearly, if the regulator is asked to make many well considered decisions early, there is a very great risk that there will be poor decisions and dissatisfaction from many quarters. This may endanger the whole concept. Time will be needed for the regulatory body to develop skills—skills that are in short supply.

1. The first action to minimize decisions has been taken, that is the 'pre-packaging' of many of the regulatory rules have been included in the re-negotiated concessions in Jakarta. The more 'the rules of the game' are included in cooperation agreements, the less discretionary decisions are needed from the regulator in early years.

2. The expertise of the regulatory body is being established. After recruitment of key positions a number of administrative matters were addressed:
   - A Work plan/budget needs refining.
   - Basic regulatory procedures have been developed and recorded.
   - Training of staff has begun.

3. Live regulatory matters have also begun to be addressed. These include:
   - Preparation of financial projections independent of the concessionaires' models.
   - The various activities included in the transition period (including the activities of the counterpart staff) have been organized and are progressing, most importantly through PERG meetings.
   - Establishment of public information centre(s).
   - Establishment of public consultant mechanisms.
   - Monitoring of the quarterly submission cycles in the transition period.
   - Preparation for the negotiation of the second five-year investment period of the concessions.

Outlook for future stable cooperation.   In the future, the JWSRB will play important roles in achieving more sustainable and strong footed cooperation. Some important aspects, such as the strategic agenda to be implemented within the first semester of 2004, already suggested by the JWSRB to all parties concerned, are explained further.

Upon completion of ICE Team's exercise on rate re-basing and presumably after a series of thorough discussions among all parties, new water charges are to be agreed as soon as possible. Subsequent strategic effort will work out further Legal Adjustments to Cooperation Agreements (LACA), which among other things will contain the result of new rate re-basing; a new procedure, mechanism and formula for automatic semi annual tariff adjustment; the rights and obligations of both parties to achieve more equality; and clarification of the function, role and accountability of the JWSRB to avoid overlapped tasks with PAM JAYA. Legal Adjustments to Cooperative Agreements was one of the requirements for the implementation of an automatic tariff adjustment in January 2006. However, according to the JWSRB, many complex issues had been only partially resolved by the end of 2005. Among these were re-basing agreements, amount of shortfall in the first five years of the Restated Cooperation Agreements and equitable treatment for all the parties. Other important issues, like equitable and fair contracts for both parties, and the functions and roles of the JWSRB, need to be immediately resolved. Socio-politically, these issues need to be resolved as soon as possible because of public interest and also due to the concern of the City Council. If not, the sustainability of the RCA, especially in terms of the implementation of an automatic tariff adjustment is likely to become more difficult to implement.

Concluding Remarks

This paper has outlined the challenges faced under Regulatory Approach to Jakarta Water Supply Cooperation. The subject is wide-ranging to the point that international experience indicates it is a profession in itself. This paper has touched on only some of the important issues. Many more will be raised in time in the Indonesian context.

The institution of the 'regulator' is fast gaining acceptance in many parts of the world as a means to improving the delivery of an infrastructure and related services. Each country has variations on a theme; each country has awarded the regulator with varying degrees of independence and discretion. While not advocating that Indonesia should blindly follow the development direction in other countries, models should be considered together with their applicability under Indonesian conditions. The concept of independent regulation has arrived; ensuring it stays and grows is the challenge. It is hoped this paper will help to provide some insights as to how that may happen.

The degree to which the regulator can balance the interests of the key stakeholders will determine the future of independent but accountable regulation.

Water Management in ArRiyadh

WALID A. ABDERRAHMAN

Introduction

The Kingdom of Saudi Arabia is situated at the furthermost part of southwestern Asia, and it occupies approximately four-fifths of the Arab Peninsula, covering a total area of 2.25 million km^2 of which about 40% are desert lands (Figure 1). In 1974 the country witnessed a major threshold increase in its oil revenues by about 20 to 40 times. The annual government revenues (mainly oil) have increased since 1974 from less than Saudi Riyals (SR) 5 billion to about SR100–220 billion. Since 1974, the country has experienced comprehensive and rapid developments socially, and in the construction, education, health, transport, industrial and agricultural sectors. The Kingdom managed to move within a limited number of years from a typical Third World country to a more advanced country with modern and complex infrastructures. The improvements in the standard of living coupled with better health care in urban and rural areas have led to rapid population growth. Large cities with a population of more than 4 million, such as ArRiyadh and

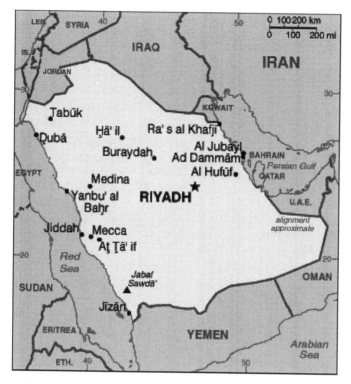

Figure 1. Location map of Saudi Arabia and ArRiyadh City. *Source:* HCDA (2005).

Jeddah, have been developed during the last decade. The rapid population growth and urbanization have resulted in a major rise in water and sanitation services, which has put immense pressure on local water authorities to satisfy these requirements, especially in an extremely arid country such as Saudi Arabia.

This paper describes the challenges in satisfying the rapidly growing demands on water and sanitation, and the learned lessons from the adopted dynamic water management approaches in such a large city. It also explains the problems that have emerged and the adopted policy solutions. Furthermore, it explains the effectiveness of water management of a new centralized water agency in satisfying the water and sanitation demands of such a fast growing city using the integrated water resources management tools.

Background Information

The city name ArRiyadh is the plural of the Arabic word 'Rawdah', which means 'Garden' or 'Meadow'. The name has been derived from natural lands, green grass cover and the fragrance of flowers. ArRiyadh was made the capital of the Kingdom of Saudi Arabia by King Abdulaziz bin Abdulrahman Al Saud in 1902 (1319/20 AH).

ArRiyadh city is located almost in the centre of Arabian Peninsula, on a sedimentary plateau approximately 640 m above sea level. Wadi Hanifa is the most prominent feature of the city that flows through from north west to south east for about 120 km. It acts as a natural drainage of rainwater for an area covering 4000 km².

The climate in ArRiyadh is hot in summer and mild in winter. The highest temperature is in summer when it can reach 42°C, and it decreases in winter to about 11°C (MAW, 1986). July is the hottest month while January is the coldest. Annual precipitation is low with an average of less than 150 mm/year.

ArRiyadh is a good example of the rapid transformation of a small and conventional Third World city with an area of less than 1 km^2 in 1918, approximately 100 km^2 in 1970 and is now a large modern city covering an area of approximately 1600 km^2, supported by complex and advanced infrastructures inhabited by a population of about 4.26 million in 2004. In less than half a century, its area has expanded over 100 times (Figures 2, 3 and 4) (Aldahmash, 2002; HCDA, 2005). Today, ArRiyadh is considered to be among the fastest growing cities in the world.

The population in the Kingdom increased by 300% between 1970 and 2004, and it is expected to double during the next 20 years. The population growth rate has ranged between approximately 3.1–3.6%, especially since 1975. The population has increased from about 4.07 and 7.7 million in 1960 and 1970 respectively, to about 10.7, 15 and 22.5 million in 1980, 1990 and 2004 respectively, and it is expected to reach about 36.4 and 41 million by 2020 and 2025 (MFNE, 2004). The urban population has increased from approximately 3.74 million or 50% of the total population in 1970 to approximately 6.4 (60%), 10.5 (70%) and 15 (72%) million in 1980, 1990 and 2004, respectively. The urban population is expected to reach about 32.8 million in 2025 or approximately 80% of the total population of the country.

The population of ArRiyadh has increased from approximately 20 000 in 1918 to 200 000 in 1960, 400 000 in 1970, 1.094 million in 1980, to 2.1 million in 1990; 4.26 million in 2004, and it is expected to reach approximately 10 million in 2020 (Figure 5) (MFNE, 2004). The rapid and huge increase in population is attributed to a high rate of population growth, and continuous movement of families from rural to urban areas for better job opportunities similar to other developed countries.

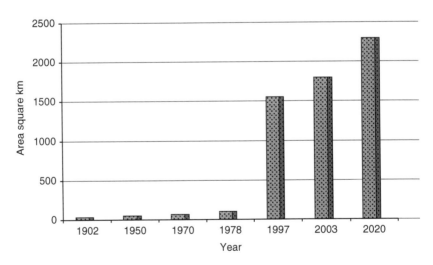

Figure 2. Growth in the area of ArRiyadh city since 1902.

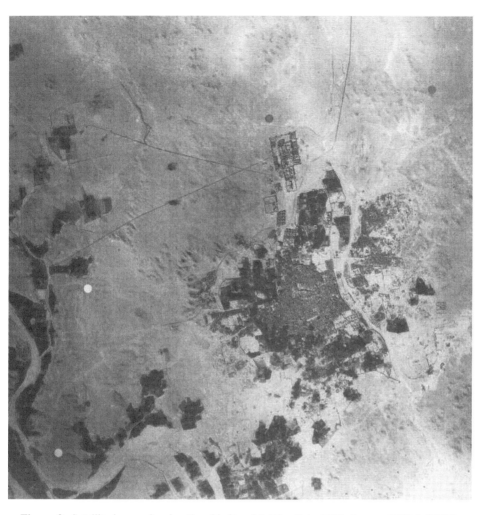

Figure 3. Satellite image showing the old city of ArRiyadh in 1975. *Source:* HCDA (2005).

Domestic Water Demands

The current national water demands for domestic, industrial and agricultural uses in the Kingdom are about 21 274 million m³ per year (m m³/year). The national agricultural water demands are 19 074 m m³/year or about 2375 litres/person/day (l/p/day). The national domestic and industrial water demands have grown from approximately 220 m m³ in 1970 to about 2030 m m³ in 2000 and approximately 2200 m m³ in 2004. These demands are expected to reach 6450 m m³ in 2020 (ESCWA, 1999; MAW, 2002; Al-Hosayyen, 2002 and personal estimations). About 86% of these demands are for domestic use and the rest are for industrial use. The average domestic water demands are about 235 l/p/day. The growing domestic demands are mainly satisfied from desalination plants and from the non-renewable groundwater resources.

In ArRiyadh city, the domestic and industrial water demands have increased from about 50 million m³/year (m m³/year) in 1970, to 80 m m³/year in 1980, to 420 m m³/year in

Figure 4. Recent satellite image for ArRiyadh City in 2003. *Source:* HCDA (2005).

1995, to 566 m m^3/year in 2000 and 579 m m^3/year in 2004, and are expected to reach 1130 m m^3/year in 2020 (Figure 6) (MOWE, 2005). The average domestic water use in the city in 2004 was about 497 m m^3/year or 320 l/p/day.

Domestic Water Supplies

The rapid expansion in the area of ArRiyadh city together with fast population growth, particularly during the last 35 years, have been the major challenges for the local Water Directory to build water supply networks and water production facilities within a limited time to meet the huge increase in water demands. Furthermore, the available groundwater resources in the aquifers near ArRiyadh cannot supply the increasing quantities and their salinity levels are not suitable for drinking purposes. The local water authorities used the following sources for supplying the city with domestic and industrial water:

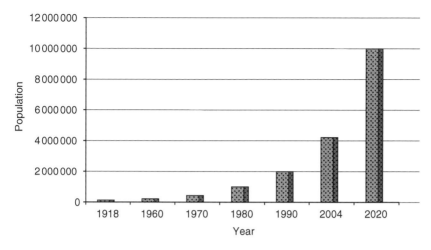

Figure 5. Growth in the population of ArRiyadh city since 1915.

- *Desalination water*: Approximately 830 000 m³/day (302 m m³/year) of desalinated water using the Multi Stage Flush System (MSF) is produced in the Al-Jubail desalination plant on the Arabian Gulf coast, and is pumped to ArRiyadh through two pipes, each with a radius of 60 inches, for a distance of 466 km. Further, expansion of the Jubail desalination plant is being considered to meet the increasing demands (MOWE, 2005 and personal communications, 2005).
- *Groundwater*: Groundwater from local aquifers around ArRiyadh is pumped and treated by the reverse osmosis desalination process (RO) to augment the desalination water from the Jubail plant. Approximately 7 m m³/year of groundwater are pumped from Wadi Nesah. Groundwater is also drawn from the Minjur aquifer near ArRiyadh, tapping two locations. About 39.4 and 31.4 m m³/year are pumped from the Bowaib and Salboukh fields respectively from

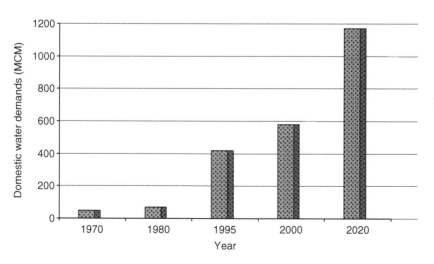

Figure 6. Growth in the domestic water demands in ArRiyadh city since 1970.

the Minjur aquifer. Further, about 73 m m^3/year is pumped from well field tapping The Wasia aquifer is situated 110 km from ArRiyadh. New groundwater pumping fields from the Umm Er Radhumah aquifer in Al Honai approximately 218 km to the east of ArRiyadh have recentsly been developed in 2005. This field has been producing about 127 m m^3/year to augment the water supplies to the city in order to satisfy the rising demands (personal communications, 2005). The total groundwater supplies to the city are now about 277 m m^3/year or about 47% of the total water supplies.

The number of house connections to the water network have increased from 83 222 in 1980 to 240 037 in 2000, 280 000 in 2002, 291 000 in 2004 and are expected to reach 499 670 in 2020 (MOWE, 2005). Currently, the water networks in the city are about 9860 km in length and about 98% of the ArRiyadh population is served by drinking water services.

The estimated costs of desalination water production, transportation and distribution are about US$1.41/m^3 at capital investments of 10%. The costs of groundwater production, transportation and distribution at capital investment costs of 10% are about US$0.42/m^3. The government has fully supported the costs of water and sanitation services in the Kingdom, including ArRiyadh city. The total cost of domestic water production and distribution, in addition to sanitation services in the Kingdom, have exceeded US$100 billion since 1975. About 20% of theses costs have been for water and sanitation services in ArRiyadh (Al-Hosayyen, 2002; Aldahmash, 2002). The domestic water tariff in the Kingdom is very low and thus does not generate enough revenues. The total annual revenues do not exceed 2.5% of annual expenditure. However, in spite of the population increase in the city to about 4.26 million in 2004, the water supplies have not significantly exceeded the 1992 water supplies level of 415 m m^3/year (1 137 804 m^3/day) which was used to supply the population of 2.29 million in 1992 (Aldahmash, 2002) (Figure 7). This is due to a lack of the huge financial resources necessary to support the execution of the required facilities within a few years, especially with the decrease in oil prices between 1995–2002. However, this problem has vanished since the rise in oil prices since 2003. It is estimated that the Kingdom requires about US$130 billion to satisfy the growing water and sanitation services in the whole country until 2022. ArRiyadh city needs about US$29 billion to meet the rising water and sanitation services for a population of about 10 million in 2022. Approximately 45% of these costs are for domestic water production and distribution facilities until 2022.

The rapid growth of residential and commercial areas of the city during a limited number of years has made it difficult for the local water directorate to provide the sanitation services similar to water supplies (Figure 7). This is because the wastewater collection networks and treatment facilities require special, more complicated and costly works. The sewerage system network covers approximately 35% of the total area of the city and about 56% of the city population benefit from it. The total length of this network in 2002 was 2322 km, and the number of users of sewerage services in the city reached 159 117 users (Aldahmash, 2002).

The wastewater is collected and treated in three wastewater treatment plants at the tertiary stage. The capacity of the three plants is approximately 403 000 m^3/day or 147 m m^3/year (MOWE, 2005). Most of the effluents are reused for restricted irrigation downstream of the city in Wadi Hanifa, and for landscape irrigation within the city. The

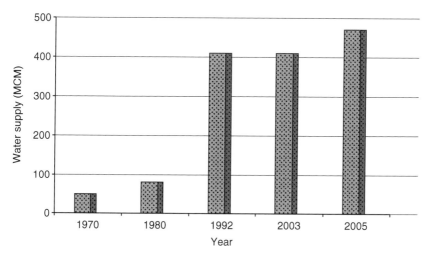

Figure 7. Growth in water suplies for ArRiyadh city since 1970.

government has fully supported the execution of the sanitation services including networks and treatment plants. It has invested about US$10 billion on wastewater collection and treatment facilities in ArRiyadh during the last three decades. However, the investments required for additional wastewater facilities are huge. This is estimated to be about US$16 billion to meet the demands for sanitation facilities until 2022. Hence, low oil prices and the lack of funding has also added to slowing the development of the sewerage system coverage, especially between 1995–2002. However, since 2003, financial support together with the rise in oil prices, have been better and the expansion of the sanitation facilities has been very effective.

Problems of Water and Sanitation Services

Various types of problems related to water and sanitation services have been experienced in ArRiyadh, especially before the establishment of the Ministry of Water and Electricity. This is due to: (1) rapid growth in the city population, area and water demands within a short space of time; (2) the high costs of water and sanitation services; (3) the high average water consumption; (4) high levels of water leakages (20–30%) from water supply networks; (5) fragmented responsibilities among various agencies which used to deal with water issues; (6) some 30% of additional water storage facilities are needed to supply water to the city during emergency situations; and (7), the focus was on water supply management and not enough on demand management and conservation. Examples of the water and sanitation problems are:

- Water shortages, especially during the peak demands in summer months, were experienced until 2004. Accordingly, water had to be distributed rotationally among the various sectors of the city. An additional 18 m m^3 is distributed by private water contractors, especially during the summer season.
- Serious disruptions of water supplies in the city, especially during emergencies, such as a breakdown in the main water transport pipeline.
- Losses of large volumes of costly desalinated water due to high levels of water

leakages from water supply networks.

- The development of a shallow water table rise problem in various parts of the city. Water leakages from the water and wastewater networks, seepage from septic tanks in the unserved parts of the city and the excess irrigation water from landscape irrigation have contributed to a shallow water table rise problem in the city. This problem has had negative impacts on roads, highways buildings and different types of facilities in the city. The economic losses as a result of this problem have been estimated to be SR885 million annually (HCDA, 2005).

- Losses of large volumes of wastewater annually which could be treated and reused.

- Possible pollution of drinking and groundwater by wastewater seepage from septic tanks, possible leakages from wastewaters networks and leakage from petrol stations and car service centres.

Water Management Approaches in ArRiyadh

Water Management Approaches Prior until 2001

Prior to the establishment of the Ministry of Water and Electricity in 2001, several governmental agencies were involved in water production, treatment and distribution, and wastewater collection and treatment. These agencies were: (1) the Saline Water Conversion Corporation (which used to be part of the Ministry of Agriculture and Water) which was responsible for desalination water production and transportation from the Jubail desalination plant, approximately 466 km from ArRiyadh on the Arabian Gulf coast; (2) ArRiyadh Water Authority (which used to be part of the Ministry of Agriculture and Water) which was responsible for groundwater production, desalination and transportation to ArRiyadh; (3) ArRiyadh Water Authority, whose name was changed to ArRiyadh Water and Wastewater Authority (which used to be part of the Ministry of Municipal and Rural Affairs), which was responsible for water distribution networks and sanitation facilities in the city; and (4) the High Commission for the Development of ArRiaydh, which has been responsible for developing policies and procedures aiming to raise the efficiency of services and facilities related to improving the living standard of citizens and their welfare. It is also responsible for the comprehensive development, the coordination and implementation of the basic infrastructure projects in the city, in addition to conducting planning works and studies.

The efforts of above agencies managed to execute the following:

- Water supply management. Each of above agencies contributed within its capacity to the development and execution of water supply plans from groundwater and seawater desalination, in addition to wastewater collection and treatment facilities. Within a limited number of years and using the available governmental funds, it was possible to satisfy most of the growing domestic water demands in addition to about 35% coverage of the city with sanitation services.

- Water conservation measures. Water conservation awareness campaigns were carried out on TV, local radio and newspapers in addition to educational presentations and pamphlets to the public.

- Implementing water leakage assessment and control programmes in several parts of the city, the fixing of water consumption meters at house levels, the adoption of a special water tariff for the city (higher than other parts of the Kingdom), the reuse of wastewater for agricultural and landscape irrigation, and developing and implementing special regulations for minimizing wastage of water.
- Implementation of capacity building programmes to the local managers, engineers and technicians on water and wastewater services.
- Controlling the shallow water table rise problem. The High Commission for the Development of ArRiaydh carried out a comprehensive study to assess the causes of the shallow water table rise problem and proposed suitable solutions to the problem during the late 1980s. The Commission supervised the successful implementation of the defined plan for controlling the problem (HCDA, 1990).

In general, the above joint efforts were successful in securing water supplies from groundwater in aquifers near the city in addition to the transportation of desalination water from the Gulf coast. The executed water production, treatment and transportation to ArRiyadh using feasible engineering technologies in addition to sanitation services are important accomplishments. However, continuous rapid urbanization and population growth have resulted in a further rise in demands for water and sanitation services in the city. Consequently, more aggressive measures had to be implemented in an integrated approach under a central national water agency, which combines all efforts for planning, development and management of water and sanitation services.

Integrated Management Approaches and Accomplishments since 2001

In July 2001, it was announced that a new Ministry had been established to take charge of the Kingdom's water affairs in an integrated approach. The Ministry of Water embraces all water-related departments at what were the Ministry of Agriculture and Water (now the Ministry of Agriculture), the Ministry of Municipal and Rural Affairs and other government agencies that previously had responsibility for water. The Ministry of Water was also responsible for water and sewage discharge affairs. The Ministry is to prepare a comprehensive plan to establish water and sewage networks throughout the Kingdom. It will also develop the country's water policies and propose new regulations to preserve water. The Ministry will develop and implement new water tariffs for all groups of beneficiaries and set out a plan to ensure the effective collection of water revenues. It will also license the digging of wells and determine their depths. The Ministry will develop a suitable framework for private investment in the sector, covering finance, implementation, operation and maintenance of water projects. In May 2003 the formation of the Ministry of Water and Electricity (MOWE) was announced, bringing together these two vital aspects of the Kingdom's economy under one ministerial roof (MOWE, 2005).

Within less than three years, the new MOWE has taken several important measures at both the national and local levels, including ArRiyadh city, to achieve its goals. These measures have included the following:

- Adoption of integrated water resources management (IWRM) as the main tool to achieve sustainability of water resources and development of the Kingdom.

IWRM is a process, which promotes the coordinated development and management of water, land and related resources in order to maximize the resultant economic and social welfare in an equitable manner without compromising the sustainability of vital ecosystems.

- Commencement of developing a National Water Resources plan and a strategy for implementation using IWRM. The national water plan is the first step toward achieving sound water management and the sustainability of water resources in ArRiyadh and at a national level.

- Implementation of more focused water demand management programme has been initiated in the cities, particularly in ArRiyadh. The plan includes a more aggressive leakage detection and control programme in the city to minimize the water losses, to increase the available water for distribution in the networks and to control its contribution to shallow water table rise. A new water tariff is under development to enhance water conservation and to recover part of the costs of water and sanitation services.

- A realistic restructuring programme has begun for the water sector to achieve more flexibility in responding to the emerging operation and management needs.

- Adoption of public–private partnership (PPP) in water and sanitation services has begun. This is to increase the water production efficiency, to decrease the costs of water and sanitation and to generate enough investments to build the required water and sanitation facilities within the next 10–15 years.

- The Ministry is reviewing, modifying and developing the organizational, institutional and legal frameworks for using the IWRMS tools and for the successful planning and operation of the water sector, including ArRiyadh city.

- The Ministry has begun the implementation of an ambitious and serious water conservation plan. This has included a comprehensive awareness campaign and an investment of approximately US$15 million on the distribution of free in-house water fixtures (by taps, showers and toilet boxes) for the reduction of water consumption. The expected results have been a saving of between 25–35% of water use in homes, including ArRiyadh city.

- Commencement of a national programme for updating information about groundwater in several main aquifer systems in the country in addition to qualitative assessments of the used groundwater qualities for domestic purposes, including ArRiyadh.

- Implementation of several projects for the renovation and building of some desalination plants, wastewater treatment plants and developing new groundwater pumping fields for the augmentation of available water supplies for domestic purposes. Among these projects is the Al Honai well field which has recently begun operating to supply ArRiyadh with about 127 m m^3/year (about 350 000 m^3/day). This project is a major step forward in solving the problem of water shortages in ArRiyadh, particularly during summer season.

- Consideration of the Aquifer Storage Recovery approach for the storage of desalinated water during the times of low water demand in confined aquifers near ArRiyadh and to be used for supplying the city during emergency conditions. Furthermore, the same approach will be tested for the storage of treated wastewater effluents and for them to be re-used for irrigation purposes in areas near the city.

The above achievements indicate clearly that one specialized water Ministry can achieve significant progress in developing and implementing efficient water management schemes to satisfy the national and local needs and to achieve the sustainability of water resources and development schemes at a national level and at large city level such as ArRiyadh.

Conclusions

The urban water management experience for satisfying water and sanitation services in a rapidly growing large city such as ArRiyadh is unique in the types of challenges. It shows that such cities require special planning and management tools in addition to quick and fast responses and actions. Lessons learned have shown that the focus on water supplies more than demand management result in a sharp rise in water demand and its costly supplies, in addition to significant losses of water from the networks and untreated wastewater from the unserved areas. Consequently, shallow water rise and its related negative impacts were experienced in different parts of the city until control solutions were defined and implemented. The fragmentation of responsibilities and actions among different agencies in the city prior to 2001 (before the establishment of the Ministry of Water and Electricity) resulted in slowing the implementation of a comprehensive water management plan using an integrated approach. Since 2001 the equal focus of the centralized water agency for the city on supply tools and demand management and conservation tools using integrated approaches has resulted in significant improvements in water and sanitation services. This has been greatly enhanced by the modifications in the organizational, institutional and capacity building frameworks by the Ministry of Water and Electricity. This proves that sound management of water and sanitation in large cities requires special vision at the city level as part of a national water strategy. Its implementation should be supervised by a specialized centralized agency as the supervising body. The availability of large financial resources are also important and are the basic elements for the timely and successful implementation of water and sanitation schemes in large cities, especially those with rapid growth and under arid conditions such as ArRiyadh city. The aridity of the region and the lack of enough quantities and suitable quality of water supplies for domestic use from conventional groundwater resources around the city have forced the local water authorities to adopt the desalination of brackish groundwater and sea water as a strategic option to augment the water resources for the city.

Acknowledgements

The author wishes to thank the Research Institute, King Fahd University of Petroleum and Minerals, Dhahran, Saudi Arabia, for the support given to complete this paper.

References

Aldahmash, M. A. (2002) View of comprehensive and strategic planning of ArRiyadh to satisfy the financial requirements for public infrastructure, in: *Proceedings of the Symposium of Infrastructure Financing and Provision, Power, Water and Sewage, January 2002* (ArRiyadh: High Commission for the Development of ArRiyadh).

Al-Hosayyen, A. (2002) Water challenge-alternatives and non-conventional solutions, in: *Proceedings of the Symposium of Infrastructure Financing and Provision, Power, Water and Sewage, January 2002* (ArRiyadh: High Commission for the Development of ArRiyadh).

Economic and Social Commission for Western Asia (ESCWA) (1999) *Current Water Progress Achieved in the Implementation of Chapter 18 of Agenda 21, with Emphasis on Water for Sustainable Agricultural Production: Case Studies* (Beirut, Lebanon: ESCWA).

Higher Commission for the Development of ArRiaydh (HCDA) (1990) *Program for the control of the shallow water table rise problem in ArRiyadh City*. Final Report, Center for Projects and Planning (ArRiyadh: Higher Commission for the Development of ArRiaydh).

Higher Commission for the Development of ArRiaydh (HCDA) (2005) Available at http://www.arriyadh.com (accessed, July 2005).

Ministry of Agriculture and Water (MAW) (1986) *Climate Atlas of Saudi Arabia* (ArRiyadh: Department of Water Resources Development, Ministry of Agriculture and Water).

Ministry of Agriculture and Water (MAW) (2002) *National Water Balance* in Arabic (ArRiyadh, Saudi Arabia: Data published by Ministry of Agriculture and Water).

Ministry of Finance and National Economy (MFNE) (2004) *Statistical Yearbook*, 20th issue (ArRiyadh: Central Department of Statistics, Ministry of Agriculture and Water).

Ministry of Planning (MOP) (2000) *Seventh Development Plan* (ArRiyadh, Saudi Arabia: Ministry of Planning).

Ministry of Water and Electricity (MOWE) (2005) Available at http://www.mow.gov.sa

Water Management in Dhaka

KHONDAKER AZHARUL HAQ

Introduction

Bangladesh is one of the most densely populated countries in the world. More than 130 million people live in an area of 147 540 km^2 and the population is increasing at a rate of around 1.6% annually. Approximately 44% of the population lives below the poverty line. The urban population is around 25%. The country is administratively divided into six divisions, 64 districts, 460 *Upazillas* (sub-districts) and 4450 unions. *Upazillas* are the lowest tier of administrative unit and the unions are the lowest tier of the democratically elected local government. Recently a new tier of local government named *Gram Sarkar* (village government) has been created at the village level with representatives nominated by the government.

The capital city Dhaka is a megacity with a population of 12.5 million (including the suburban town of Narayanganj) that is growing at an annual rate of around 5%. The population of other cities and towns are also growing fast and it is expected that nearly 50% of the population of the country will be living in urban areas by the year 2020. Urban areas are contributing an estimated 40% to the GDP. While catalyzing economic growth, rapid urbanization also places considerable burden on urban areas, both in terms of additional stress on an inadequate urban infrastructure and services and the exacerbation of already poor environmental conditions.

Government agencies, autonomous bodies, local government institutions such as specialized water utilities, municipalities and city corporations are responsible for providing the water supply and sanitation services. The Local Government Division (LGD) under the Ministry of Local Government, Rural Development and Cooperatives (LGRD&C) is the apex body that oversees and controls the activities of these organizations.

Urban Water Supply and Sanitation

Water supply and sanitation is basically a public sector responsibility. The Department of Public Health Engineering (DPHE), a government agency, is responsible for water supply and sanitation for both urban and rural areas of the country, except for the capital city of Dhaka and the port city of Chittagong. In order to meet the growing demands for water supply and sanitation services of the two cities, two autonomous organizations, Dhaka Water Supply and Sewerage Authority (DWASA) and Chittagong Water Supply and Sewerage Authority (CWASA), were created in 1963 under Water Supply and Sewerage Ordinance (East Pakistan Ordinance No. XIX of 1963). Since then, the DPHE/municipalities have been given the responsibility of providing water supply and sanitation services to the rest of the urban areas and the entire rural areas of the country. DPHE'S mandate in terms of urban water supply and sanitation include construction, improvement, modernization and expansion of the municipalities and other growth centres. After completion of the projects, the facilities are transferred to the appropriate municipality/local government authority for operation and maintenance with their own funds and funds collected for the services provided.

In 1971, when the country became independent, a piped water supply existed in limited areas of 40 towns. Piped water supply was supplemented by manually operated tubewells. Since the old municipal water supply system could not cope with the growing demand of fast increasing urban population, different projects were implemented with the assistance of the donor agencies, increasing the water supply coverage substantially. A number of new towns were also brought under the piped water supply system. At present there are 237 municipalities in the country and the piped water supply system has been developed, partially or fully, in 101 of them. Water supply coverage is about 50% in the urban areas, except for Dhaka and Chittagong where the coverage is 72% and 45% respectively (Andrews & Yniguez, 2003).

Sanitation coverage in Dhaka city is around 70%, of which nearly 30% may be assigned to the water-borne piped sewerage. The rest of the sewerage is handled through conventional septic tanks. Nearly one-third of the domestic effluents are released directly into the environment without any treatment, which is the major source of pollution of the surface water bodies located in and around the city. Some of these water bodies also serve as raw water sources for the surface water treatment plants.

Policies and Strategies for Urban Water Supply and Sanitation

The Bangladesh Government formulated a national policy for water supply and sanitation in 1998 (GOB, 1998). The broad objectives of this policy are to improve the standard of public health and the overall quality of the environment. Amongst the important goals are the following:

- Making safe drinking water available to each household in urban areas at affordable costs.
- Ensuring access to sanitary facilities to all urban households through appropriate technical options ranging from septic tanks to a pipe-borne sewerage system.
- Providing adequate water supply and sanitation facilities to schools, bus stations and other important public places, and also to densely populated poor communities not having sufficient space for the provision of individual household facilities.
- Ensuring potable water supplied meets required quality standards.
- Removing arsenic from drinking water, and supplying arsenic-free water from alternate sources, wherever needed.
- Disposing of solid and liquid wastes in environmentally safe ways, and removing storm runoff.

Policy Highlights: Urban Water Supply

WASAs have been made responsible for the sustainable water supply in the metropolitan areas of Dhaka and Chittagong. In other urban areas, the municipalities, with the help of DPHE, are responsible for providing services. In order to meet their mandates, WASAs and municipalities will have to improve their current operational efficiencies very significantly, including financial management. The billing and collection targets need to be raised to 90% and 80% respectively. Municipalities and WASAs must also take adequate measures to prevent wastages of water. In addition, they should take the necessary steps to raise public awareness to prevent misuse of water. Appropriate measures need to be taken to reduce unaccounted-for-water (UFW), from the current 50% to at least 30%. Dhaka and Chittagong WASAs must also reduce their shares of unaccounted-for-water from their present levels, which are very high as per international standards. In future, government's development grants to improve operational efficiencies will consider the following factors:

- Water supply coverage in terms of area and population;
- Level of amount of unaccounted-for-water; and
- Increase the revenues of the institutions.

Private sector participation will be promoted through BOO/BOT (Build Own and Operate/Build Operate and Transfer), outsourcing specific operations, and through other forms of acceptable arrangements. For this purpose, opportunities will be created for involving the private sector initially in revenue billing and collection. At a later date the government will prepare comprehensive guidelines for private sector participation in the water sector.

Urban Sanitation

Under the national policy, the sanitation system should become self-sufficient and self-sustaining. Sanitary facilities in every household will be promoted. Public and community facilities will be constructed by the city corporations/municipalities, and will be leased out to the private sector for maintenance.

The city corporations or municipalities are responsible for the collection, transportation, disposal and overall management of solid wastes. These organizations may transfer, where feasible, the responsibility of collection, removal and management of solid wastes to the private sector. Where WASAs exit, they will be responsible for sewerage and stormwater drainage systems.

Behavioural changes in user communities will be brought about through social mobilization and hygiene education, in cooperation with the Ministries of Health, Education, Social Welfare, Information and Women & Children Affairs, DPHE, NGOs, CBOs (Community Based Organizations), local government bodies and other related agencies. Measures will be taken to recycle waste materials as much as possible and to prevent the contamination of groundwater by sewerage and drainage discharges.

With regard to institutional arrangements in the water supply and sanitation sector, the Local Government Division will continue to remain responsible for overall planning, identification of investment projects and coordination of activities of agencies under it. However, each of the relevant organizations will be responsible for its own activities. The Local Government Division will constitute a forum with representatives from relevant organizations to coordinate, monitor and evaluate the activities of the sector and to determine the future work programmes.

Institutional Arrangements for Urban Water Supply and Sanitation

Relevant WASAs will be responsible for sanitation in the Dhaka and Chittagong city areas. The involvement of the private sector in these activities will be explored, examined and encouraged. In other urban areas the Department of Public Health Engineering, either solely or jointly with municipalities, is responsible for such services.

Local government bodies such as *Zilla Parisad* (district council), *Upazilla Parisad* (sub-district council), *Union Parishad* (union council) and *Gram Parishad* (village council) will be gradually provided with more opportunities to contribute in the activities of the sub-sectors. A congenial atmosphere will be created and the necessary support provided to facilitate increased participation of the private sector, NGOs and CBOs in the activities of the sector both in the rural and urban areas.

Monitoring of water quality for the purpose of ensuring compliance with acceptable standards will be done through DPHE, WASA, the Department of Environment (DOE), Atomic Energy Commission (AEC), Bangladesh Council for Scientific and Industrial Research (BCSIR) laboratories and the Bangladesh University of Engineering and Technology (BUET). Reports will have to be submitted to the government at prescribed intervals.

WASAs and relevant agencies will support and promote any collective initiative in slums and squatter settlements in accessing a water supply service on payment. NGOs will play appropriate role in undertaking promotional activities. Private sector and NGOs will be encouraged to invest in the manufacture, sale and distribution of different types of tubewells and accessories for such areas. They will also be encouraged to participate in the installation of a piped water supply system where feasible.

All relevant organizations will emphasize a reduction of over-dependence on groundwater and an increased use of surface water. In order to make the water supply system sustainable, water should be supplied at least cost. Educational and religious institutions will be provided with water as per rules prescribed by the government from time to time. In future the water tariff will be determined on the basis of the cost of water production, operation and maintenance, administrative overheads and depreciation.

Dhaka Water Supply and Sewerage Authority (DWASA)

As has been mentioned previously, DWASA was created in 1963 by an ordinance with modest human resource and logistical support with the mission to "Provide potable water and sanitation services to the city dwellers at an affordable price" with a vision to ensure "100% water supply coverage by the year 2005 and 80% sanitation coverage by 2020". Over the last 40 years it has grown into a major water utility in the region serving more than 12.5 million people in a service area covering $350\,km^2$. The organization is responsible for supplying potable water and the disposal of domestic sewerage and stormwater from the city of Dhaka, the capital city of Bangladesh and Narayanganj, a suburban business centre 12 km from Dhaka. The current population of Dhaka is estimated to be around 12 million and that of Narayanganj about 500 000. As mentioned previously, the service area population is increasing by about 5% per annum, which is nearly 400% higher than the national average of 1.6%.

For operational purposes, the entire service area for DWASA has been divided into seven zones, six in Dhaka and one in Narayanganj. Each zone has two major responsibilities: (1) the provision of a water supply and domestic sanitation services; and (2) billing and collection of revenue. Stormwater drainage is handled by a separate office headed by a senior engineer.

Governance and Organization

DWASA is governed by a Board of Directors consisting of 13 members drawn from different professional bodies and civic society. The Board is responsible for policy matters relating to corporate planning, tariff setting, deciding on the organizational structure and staffing, appointing senior management, deciding on their remuneration etc. The Chairman of the Board and 10 out of 13 members are appointed the Ministry of LGRD&C from outside the government. The Managing Director and three Deputy Managing Directors are appointed by the Board with prior approval from the government

The management is headed by a Managing Director, who is the Chief Executive Officer (CEO). He is assisted by three Deputy Managing Directors, one each for Administration and Finance (A&F), Operation, Maintenance and Logistics (O&M) and Research, Planning and Development (RPD). Except for the DMDs the Managing Director is responsible for recruitment, promotion, transfer and general administration of all other staff of the authority.

The three mainstream activities of the authority are:

- Administration and finance;
- Project preparation and implementation; and
- Operation and maintenance of water supply and sanitation infrastructures.

Staffing and Human Resources

DWASA employs some 3783 employees. The staff breakdown is as follows: technical services 2872; administrative services 382; financial services 529.

In 1996 the staff connection ratio was 18.5 per 1000 connections (McIntosh & Yniguez, 1997) which was significantly reduced to 11.6 per 1000 connections in 2001 (Andrews & Yniguez, 2003) but increased to 14.2 at the end of 2003 (WASA, MIR, December, 2003). The comparable numbers from Colombo, Delhi, Ho Chi Minh City, Karachi and Manila are 7.6, 19.9, 3.5, 6.4, 15.2, 4.4 respectively (Andrews & Yniguez, 2003). There is one significant difference in the water supply between Dhaka and other large water utilities. In Dhaka water is supplied from nearly 400 supply points whereas in most of the cities in the world water it is supplied from less than 20 supply points. Even then the staff connection ratio should be brought down to around 10 to make the utility leaner and more efficient.

Water and Wastewater Management

As mentioned previously, DWASA is responsible for providing a water supply and sanitation service (domestic sewerage and stormwater disposal). The following section describes the present status and future strategies for expanding and improving water and wastewater management to meet the fast increasing needs of the city dwellers.

Water Supply Management

At present demand for water is around 1760 MLD at 160 l/per capita. Against this demand DWASA has a production potential of 1600 MLD, but actual production ranges from 1400 to 1500 MLD. The water supply system is groundwater based and 82% of the supply is abstracted from underground aquifers through 390 production wells. The rest 18% is derived from three surface water treatment plants of which the Saidabad Water Treatment Plant (SWTP) is the largest. The plant was commissioned in July 2002 and has a capacity of 225 MLD. Before SWTP was built the dependence on groundwater was nearly 97% (DWASA, 1999/2000).

The first modern water supply system for the city was commissioned on 6 August 1874. The plant was built through a grant from Nobab Khajeh Abdool Gunny, the then ruler of Dhaka under the British Raj. The event was considered so important that on that day the Viceroy and Governor General of India, the Right Honorable Thomas George Baring, graced the occasion with his presence. The Lt. Governor of Bengal, Sir Richard Temple, was also present.

As of December 2003, the water supply system consisted of 390 production wells delivering 1284 MLD and three surface water treatment plants producing 255 MLD 2400 km of water distribution lines of different diameters, 216 800 water connections and 38 overhead water reservoirs (DWASA, MIR, December 2003). The water supply infrastructure of the city is shown in Figure 1.

The service area population, water demand, actual water supply and deficit are shown in Table 1. It can be seen from the Table that DWASA started its operation in 1963 with a service area population of 0.085 million and a modest deficit of 13% of water. The gap between demand and supply continued to increase for the next 33 years and peaked at 49% in 1990. From 1996 the trend was reversed and the deficit started to decline. By 2003 the

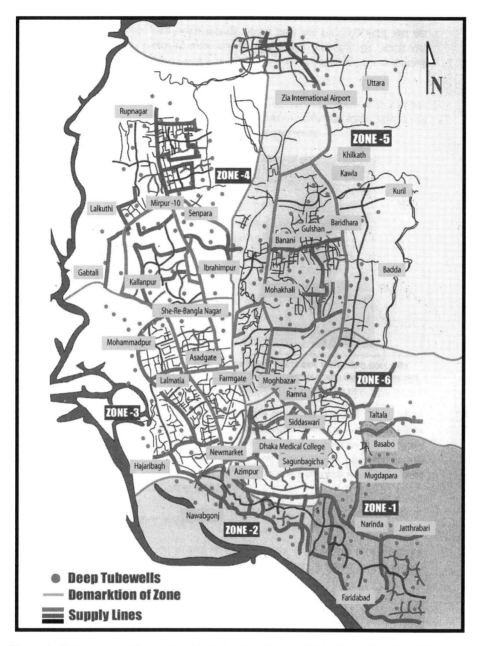

Figure 1. Water supply infrastructure of Dhaka City. *Source:* Dhaka Water Supply and Sewerage Authority.

deficit between the demand and supply was at its lowest during the entire history of DWASA except for the inception year of 1963. This was possibly due to the construction of a large number of deep tubewells and the commissioning of a major surface water

Table 1. Population, water demand, supply and deficit

Year	Population (millions)	Water demand (MLD)	Water supply (MLD)	Deficit (%)
1963	0.85	150	130	13
1970	1.46	260	180	30
1980	3.03	550	300	45
1990	5.56	1000	510	49
1996	7.55	1300	810	38
1997	8.00	1350	870	36
1998	8.50	1400	930	34
1999	9.00	1440	1070	26
2000	9.50	1550	1130	25
2001	10.00	1600	1220	24
2002	10.50	1680	1300	23
2003	11.00	1760	1400	20

Source: DWASA Annual Report (2001); DWASA, MIR, December (2003).

treatment plant with a capacity of 225 MLD in mid-2002. The plant, Saidabad Water Treatment Plant (SWTP), was jointly funded by the Government of Bangladesh (GOB), the World Bank (WB) and the Government of France (GOF), costing around US$170 million. Construction work on the plant was started in October 1997 and it was completed and put into operation in July 2002.

The deficit between demand and supply is expected to widen again from 2004, but at a much faster rate due to the inability of the DWASA to secure the necessary funding for the construction of two more surface water treatment plants. At the same time, the construction of new replacements for the old deep tubewells will also not produce the designed discharge as the groundwater reservoir of both the primary and secondary aquifers has nearly been exhausted. Although a new initiative of exploiting deep aquifers (more than 300 m depth) has yielded some promising results in a particular area of the city, any large-scale development of the deep aquifer will have to be carefully evaluated for its technical, economic and environmental viability.

Meeting the water demands of the city has the double problems of maintaining a 6% growth rate and at the same time reducing the dependence on groundwater. The problems of quality of both the surface and groundwater have also added a new dimension to the water supply scenario.

It should be noted that the first modern water supply system for the city was a surface water based system built more than a century and a quarter ago. At the time, that was probably the best system to use. However, due to various reasons, including an adequate availability of groundwater, cheaper construction costs, shorter times required for construction, the availability of local expertise for construction, all these factors have influenced the adoption of a groundwater based water supply system which has now become unsustainable.

Available data indicate that the groundwater table is falling at an alarming rate of over 2 m per year and the groundwater is being mined in significant amounts. Consequently, the discharge of the tubewells is being reduced at a much faster rate, resulting in higher costs for pumping and drying up of the boreholes. The life of boreholes has been reduced from 20 years in late 1960s and early 1970s to less than five years at present. Investment in new deep wells is fast becoming economically unattractive.

The master plan for water supply prepared in early 1990s and the water resource management study completed in 1998 (WSP International, 1998) have rightly emphasized the balanced development of surface and groundwater. As a part of the plan, the first large surface water treatment plant with a capacity of 225 MLD has been constructed and commissioned at Saidabad. The plan calls for building two more treatment plants in Saidabad, tripling the capacity. As a first step adequate land has already been acquired for the construction of all three treatment plants. For the second treatment plant Danish International Development Assistance (DANIDA) expressed an interest in financing the project and substantial progress was also made in this regard. However, the deal has not yet been finalized because DANIDA have certain conditions that appear not to have been adequately addressed by the GOB. For funding the third treatment plant with a capacity of 450 MLD the Swedish International Development Agency (SIDA) was approached. Several studies, including a feasibility study, were conducted, but due to severe reservations about the fast deteriorating water quality of the surrounding rivers the request has been put on hold.

Studies show that the groundwater resources within the greater Dhaka area are inadequate to meet even the current needs for water supply in Dhaka (WSP International, 1998). Groundwater extraction, particularly in the central, old city area of Dhaka is already beyond the sustainable limit. This implies that groundwater extraction must be reduced in this area to allow the groundwater table to stabilize and recharge to an acceptable level. In order for this to be viable, the balance of demand must first be supplied from surface water resources. Conjunctive use of surface water resources would allow the recovery of the groundwater table. In many places in the DWASA service area there is no scope for the installation of additional tubewells, since they are already too closely spaced for efficient operation.

In addition to the 390 deep tubewells (average capacity 3000 lpm) operated by the DWASA, there are more than 1000 wells of various capacities operating in the private sector. Serious concerns were raised about the possibility of land subsidence in the city due to severe mining of groundwater, as happened in Bangkok and Mexico City. These concerns were addressed by two studies. The first was conducted in 1991 by Sir McDonald and Partners in association with BUET (Sir McDonald and Partners, 1991) and which was updated in 2001 by BUET (BUET, 2000). Both studies indicated that the possibility of land subsidence was very remote due to the unique geology of the Dhaka soil. This was a welcome relief. However, both studies confirmed that the groundwater resources were depleting fast and alternate sources must be identified to meet future needs.

There is absolutely no question that conjunctive use of ground and surface water will be the corner stone for developing a sustainable water management plan for Dhaka city. The ultimate goal will be to reduce the contribution of groundwater in the total water supply from 82% to around 50% and the share of surface water to be increased from 18% to around 50%. The first part is easy, but the management of surface water is going to be an entirely different. First, building a surface water treatment plant has a much longer gestation period, second, the initial capital cost is high, third most of the equipment and technical knowledge is not available locally, and finally, the surrounding rivers are so polluted that the conventional treatment methods may not be able to purify their water for human consumption.

Quality of untreated water.   The quality of ground and surface water is being regularly monitored both by DWASA and other agencies as part of a long-term water management plan. The quality of groundwater appears to be better than surface water, except for a build-up of some inorganic elements such as chromium and lead (but they are well within the acceptable limits) and the presence of bacteria (e.g. E-coli has been detected in certain areas). Although arsenic contamination of groundwater is very widespread in the country, the groundwater in Dhaka is totally arsenic-free, which is welcome news for the city. However, the major problem continues to be the quality of the fast deteriorating surface water.

The water bodies in and around the city have become increasingly polluted as urban/industrial development has intensified. The rivers that surround Dhaka and the numerous small and large closed water bodies are the ultimate destination of all the wastewater discharge in the city. The inadequate sewage and storm drainage system, weak and often non-existent environment protection enforcement, poor solid waste management and monsoon weather patterns have all contributed to a poor living environment. The prevention of water pollution control is minimal, especially in the slum areas. Water is not usually treated at the source of pollution and is simply discharged into the various available water bodies. People with the lowest incomes then use it with no regard for the health risks.

Most of the water bodies are currently receiving untreated wastewater from domestic, industrial and commercial establishments due to the insufficient provision of sewerage and drainage facilities. The illegal dumping of solid wastes is another important reason for water pollution and environmental degradation.

Tejgaon (located in the centre of the city) and Tongi (located at the northern end of city) industrial estates are significant and complex sources of industrial pollution. Factories producing detergents, dry cell batteries, pharmaceuticals, chemical products and textile dying are all discharging untreated wastewater into the rivers and other surrounding water bodies. Another important source of pollution is a cluster of tanneries, mostly small scale, located at the south-western corner of the city on the bank of the river Buriganga. Wastewater from these tanneries, that includes hexavalant chromium (Cr^6), is discharged directly in the rivers.

Water quality results obtained so far from various sources confirm that the only relatively unpolluted sources of surface water suitable for the water supply are the upper part of river Turag above Gabtali and the river Lakhya upstream of Demra. These are the only surface waters that currently comply with the required water quality standard in the dry season (WSP International, 1998). During the last six years the water quality of these rivers has further deteriorated and the above statement may no longer be valid.

Quality of supplied water. At source, water produced by DWASA conforms to international standards, but significant contamination occurs between the source and the consumer's tap. There are two major areas where contamination occurs: (1) in the distribution system; and (2) in the consumer's underground and overhead reservoirs. It is estimated that 30% of contamination occurs in the distribution system and the remaining 70% in the consumer's reservoirs. Although consumers are regularly advised to clean their reservoirs properly they seldom do so. In the distribution system contamination occurs from clay and bacteria entering through leaking joints and holes in the pipes, especially the old ones. However, unauthorized human intervention is also a major cause of contamination. While making illegal connections consumers employ unskilled labourers to do the job and in the majority of the cases those labourers are unable to make leak-proof connections, thus exposing the area to potential contamination. Due to such contamination, the incidence of water borne diseases like diarrhoea, hepatitis, etc are high among the city dwellers. In addition, leaking pipes and joints also contribute significantly to system loss.

To overcome this these two problems, a comprehensive plan for the replacement of the old and leaking pipes should be initiated immediately. At the same time, consumers must be made aware of the adverse effects of illegal connections on water quality and that they should clean their reservoirs properly and at regular intervals.

Wastewater Management

Dhaka WASA manages two types of wastewater, domestic effluent and stormwater.

Domestic effluent management. Improved sanitation services for the disposal of domestic sewerage started back in 1923. Since then, the system has been upgraded and expanded in phases by constructing the appropriate infrastructure. At present, about 30% of the service area is covered by a piped borne sewerage system. The existing infrastructure for the management of domestic sewers consists of 1 sewage treatment plant with a capacity of 120 000 m^3/day; 49 000 domestic sewer connections; 778 km of sewer line of different diameter, 20-sewer lift stations, 1 central pump station and 3 truck interceptor sewers (Figure 2). The topography of the service area is relatively flat and the effluent cannot be transported long distances by gravity. Therefore, the system is a lift-cum-gravity type where initially pipes are installed at a depth of about 1 m from the ground surface and carried to a depth of 6 m at 0.66% slope accumulating the effluent in a sump. From the sump the effluent is lifted by pumps and discharged to an intake at 1 m depth from the ground surface and the process continues till the effluent reaches the treatment plant. After treatment the treated effluent is released to the nearby river Buriganga. The treatment plant was designed to receiving an incoming effluent having a BOD (Biological Oxygen Demand) load of 200 mg/l and SS (Suspended Solid) also of 200 mg/l and after treatment the effluent having 50 mg/l and 60 mg/l of BOD and SS respectively is released to the environment.

At present, Dhaka does not have adequate sewerage and sanitation system. The waterborne sewerage system only covers about 30% of the total population of Dhaka (or 25% of the DWASA service area). It is estimated that another 30% of the population uses approximately 50 000 septic tanks and another 15% have access to bucket and pit latrines. The remaining population does not have any form of acceptable sanitary disposal system. Much of the sewerage collection and conveyance system is in a poor condition due to the lack of a proper repair and maintenance programme, which leads to sewage overflow into storm drains during the rainy season. This results in unsanitary conditions that are aggravated by the lack of a well-functioning sullage drainage system. Under an ADB financed project, DWASA has installed a 'small bore sewerage system' in the Mirpur area. Other programmes include the rehabilitation of the Pagla sewer treatment plant, financed by JICA in the early 1990s and a sanitation component in IDA's Urban Development Project (Credit 1930-BD) involving the construction of latrines in the slum areas.

As a part of the Fourth Dhaka Water Supply Project funded by IDA, there is a comprehensive long-term programme to address sanitation and sewerage issues in south Dhaka, including the execution of priority investments for low cost sanitation and major rehabilitation of the existing sewerage system. The cost of this work has been estimated at around US$1.2 billion (Burger, 1998). To bring the unserved area (North Dhaka) under a pipe borne sewerage system, a master plan was prepared with assistance from JICA (JICA,

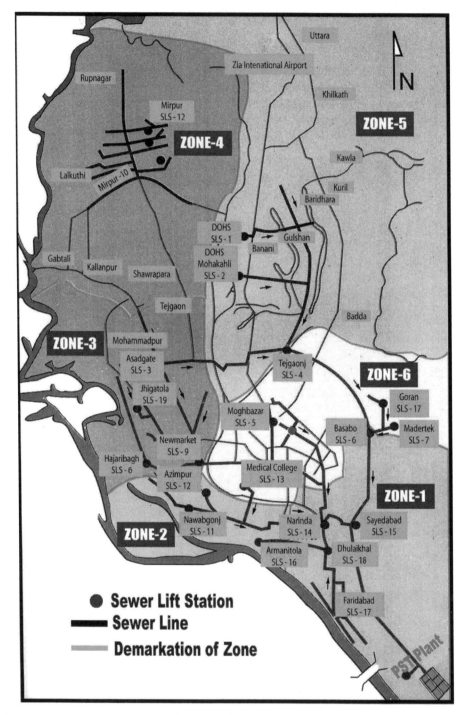

Figure 2. Domestic sewer disposal system of Dhaka City. *Source:* Dhaka Water Supply and Sewerage Authority.

1998). The study indicated that to bring this new area of approximately 80 sq km under a pipe borne sewerage system the cost would be around US$500 million.

Stormwater management. Before the 1940s Dhaka city had a natural drainage system consisting of 24 natural canals and a large area of natural wetlands that worked as retention ponds. However, as the city began to expand most of the canals were illegally occupied and the wetlands were developed as real estate. This resulted in the destruction of the natural drainage system and flooding became an unwanted way of life for the city dwellers.

The building of an infrastructure to improve stormwater drainage was begun in 1946 by the DPHE. The responsibility of removing floodwater from the city was transferred from the DPHE to DWASA in 1989. Disposal of stormwater, especially during monsoons, is a formidable task. Annual rainfall in Dhaka city averages 2000 mm. The influence of monsoon in Dhaka is relatively mild compared to the coastal area and north–eastern hilly areas of the country. Ninety per cent of this rain falls between the months of June and October. Considering the service area of 350 km^2, the city receives a total of 700 million m^3 of rainwater annually which has to be properly disposed of to prevent the city from becoming flooded for prolonged periods of time. Major drainage improvement infrastructures include three pumping stations with a combined capacity of 52 cumec, 7 km box culvert, 205 km storm drains (buried pipes of diameter ranging from 450 mm to 3000 mm) and 57 km of open canals (Figure 3). These infrastructures have proved to be inadequate to rid the city of floodwater because many areas of the city experience severe and prolonged flooding, especially during heavy showers in the monsoon season.

Major contributors to this situation are:

- an almost totally unplanned and fast-growing city;
- a reduction in the infiltration area due to the construction of roads and buildings;
- an inadequate infrastructure for the removal of stormwater;
- the disposal of solid waste in the pipes, drains and box culverts;
- inadequate funding for the development of infrastructure; and
- the closure of natural drainage canals.

In order to improve stormwater management, the following measures need to be taken immediately:

(1) Because DWASA cannot collect any direct fee for the service (unlike for water supply and sanitation), the responsibility may be handed over to the city corporation who in turn can build in the cost with a property tax. Stormwater drainage is not a conventional responsibility of the water utilities.

(2) In future the city must be developed with proper plans, providing adequate facilities for stormwater drainage and retaining all the natural drainage canals in their present form, and where required reclaiming those that have been occupied illegally. It may be mentioned here that in 1995 with the help of UNDP the RAJUK (Dhaka Improvement Trust) has developed a master plan for the future expansion of the city. The plan is called Dhaka Metropolitan Development Plan (DMDP). This plan should be strictly enforced.

(3) Natural wetlands should be preserved.

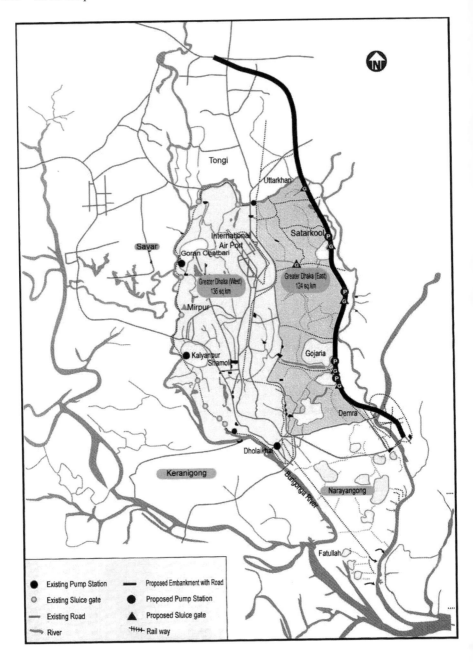

Figure 3. Stormwater drainage infrastructure of Dhaka City. *Source:* Dhaka Water Supply and Sewerage Authority.

(4)   Adequate funding should be provided to implement the master plan prepared with assistance from JICA in the early 1990s for drainage improvement and flood control of Dhaka City.

(5)   An aggressive awareness campaign should be launched to educate city dwellers to dispose of their solid waste properly instead of dumping it into drainage infrastructures.

Improving Water Management

The above discussion clearly indicates that water and wastewater management in Dhaka is well below the requirements of a megacity. On the one hand there has been a perennial gap between demand and supply and on the other hand the quality of the supplied water has not been able to satisfy the consumers. Management of wastewater, especially domestic sewage has attained alarming proportions due to a major breakdown of the existing system and the absence of a water borne sewerage system in the largest part of the city. The release of untreated sewage in the environment has literally turned the city into a cesspool. Drainage of stormwater is also unsatisfactory, which results in prolonged flooding in large parts of the city, especially during the monsoon.

Although recommendations have been provided in each section to improve water and wastewater management, the following sub-sections provide additional recommendations to improve the overall water management of the city.

Improving Governance and Organization

The World Bank, which was the major donor of the Fourth Dhaka Water Supply Project, recommended the adoption of certain changes in governance and management of DWASA before the credit was made effective. Among others, these first included the replacement of the then internal Board consisting of the Chairman of DWASA (CEO) as the Chairman of the Board and Member (Administration), Member (Finance) and Member (Engineering) as Directors of the Board by a seven-member external Board with members drawn mostly from professional groups and civic society. Later the size of the Board was increased to 13 when the WASA ordinance was enacted into law by Parliament in late 1996. Near doubling the size has made the Board too big and some times unwieldy. It is therefore recommended that the Board should be reconstituted as a seven-member Board. At present the Board is involved in many issues including procurement which are management issues and hence should be left to the management. A clear delineation of power between the management and the Board should be made by revising the relevant sections of the law.

Second, in the revised organizational structure the Chairman was redesignated as Managing Director and the Members as Deputy Managing Directors (with some changes in responsibility). The World Bank also recommended that unlike the previous practice of bringing senior management from the government, these positions should be filled from the private sector through open competition with a first-rate remuneration package to attract competent professionals. The first recruitment of the MD and DMDs was made from the private sector following the recommended procedure. However, after the WB financed project was completed a mixed system of secondment from the government as well as recruitment of personnel from government department/parastatal bodies has begun. It is recommended that the senior management should be recruited from the private sector through open competition.

Finally, the 1996 Act was formulated quickly to meet the World Bank deadline for credit effectiveness. As a result there were certain inconsistencies in the Act that should now be corrected.

Rainwater Harvesting for Improved Water Management

An attempt to shift the dependence from groundwater to all forms of surface water has been highlighted by rainwater harvesting in Dhaka. In a pioneering effort, Dhaka WASA has started to collect rainwater from the rooftop of its administrative building. In 2002 Dhaka WASA collected and utilized 11.5 million litres of rainwater. This water has been supplied for non-potable uses thereby reducing the building's water demand considerably from other sources such as surface and groundwater. Plans are in place to set up a rainwater collection system in government buildings, semi-government buildings and buildings of autonomous bodies. Dhaka has an annual average rainfall of 2 m. If more buildings could be brought into such a programme it will not only greatly reduce demands on the conventional water sources, but also significantly reduce run-off, thus reducing flooding problems. In case of a severe storm, part of the run-off can be used for induced groundwater recharge as has been done in DWASA's Rooftop Rainwater Harvesting Project. Rainwater has the added advantage that it is free from arsenic. It is, therefore, highly recommended that use of rainwater be integrated as a component in the conjunctive water management plan.

Major Constraints to Improved Service Delivery

In Bangladesh the urban water supply system is characterized by an inadequate supply, high unaccounted for water (UFW) and insufficient revenue collection that in turn give consumers a lack of confidence, and DWASA is no exception. Some major constraints have been identified as:

(1) *Inadequate tariff structure:* DWASA barely meets its operating and maintenance costs of the water supply and sanitation system. The tariff structure is designed in such a way that it cannot recover the full cost of water production. For example, the cost of purifying $1 \, m^3$ of water at the SWTP is US\$0.13 whereas DWASA charges its customer US\$0.08. Raising tariffs is not popular with the consumers and as such a tariff structure in all urban areas including DWASA does not permit satisfactory revenue generation.

(2) *Huge non-revenue water:* Unauthorized connections, leakages in the system, lack of proper operation and maintenance, absence of 100% metering system, unauthorized connection and under billing, etc, lead to a huge amount of unaccounted for water and loss of revenue.

(3) *Lack of authority:* As the government provides funds, either from its own resources or donor assistance for projects, it tries to maintain its control on the DWASA. Although it is felt that the existing tariff structure should be revised and updated in order to make the water supply and sanitation sector sustainable,

present laws and authorities do not allow them to revise the same according to their requirement.

(4) *Lack of commitment:* Lack of commitment for developing a sustainable water supply and sanitation sector has been observed. Officers and employees of the DWASA also lack commitment. Existing laws are also not fully enforced or compromised for non-payment of bills or for irregular connections.

The above factors are linked to one another, each contributing to the other, therefore creating a vicious circle trapping the sustainability issue.

Formulation of Sector Development Framework

Effective sector coordination is absent in the water supply and sanitation sector. Duplication of activities and mutually contradictory strategies have frustrated development of the sector. The National Policy for Safe Drinking Water Supply and Sanitation 1998 highlighted the need for effective sector coordination. The Unit for Policy Implementation (UPI) of the Local Government Division of the M/OLGRD &C has taken the lead in involving the stakeholders in a consultative process in the formulation of Sector Development Framework (SDF). The SDF will provide the guiding framework for policy implementation and sector development. It will plan, coordinate and monitor all sector development activities in line with the national policy. SDF should be the basis of all activities in the sector. New initiatives, donor financed or government funded, should conform to the SDF.

Addressing Local Governance, Poverty and Gender Issues

Local governance in Bangladesh lacks both capacity and resources to shoulder the responsibility of maintaining sustainable urban water services. Strengthening of local governance should be targeted towards creating awareness and ensuring service. Although local government institutions are responsible for water supply and sanitation services at present, their financial resources and management capacity are inadequate. Effective decentralization is lacking. Reforms in policy and strategy may be necessary to decentralize service delivery and to boost the capacity of local government to takeover their responsibility.

The urban poor are contributing significantly to the economy of the country. They live in slums where the environmental conditions are deplorable. Without formal recognition, they remain outside the reach of the public water supply and sanitation services. Dhaka WASA, in collaboration with local NGOs and Water Aid UK, has initiated a pilot programme to supply piped water to the slum areas of the city. Initial results have been very encouraging and a massive expansion of the programme has been planned with assistance from DFID. Policy guidelines should be framed for slums to provide basic water supply and sanitation services. Public private participation between municipalities and entrepreneurs could be encouraged for better solid waste management. The necessary structural, financial and legal reforms must be pursued to allow utilities to deliver services efficiently in collaboration with private partners. Women shoulder more responsibilities related to water and sanitation, but often have less control over access to resources. Attempts should be made to remove gender-based inequalities in access, resources and responsibilities.

Increasing Private Sector Participation

Water services have traditionally been provided within the public sector almost all over the world for social, economic and political reasons. However, in many places public services are not highly regarded, suffer from under-investment, overstaffing, low pay, limited availability of technical equipment etc. In order to rectify the deficiencies, the government is now trying to attract the private sector to the utility services. Of the various options for private sector participation, DWASA contracted out the billing and collection of three revenue zones of DWASA Employees Consumers Supplies Cooperative Society Ltd. (ECSCSL). A marked improvement in the volume of billing and collection of revenues has been observed. The programme has been found to be cost-effective and has resulted in some gains to DWASA. DWASA has plans to contract-out the remaining zones to private operators in phases. The programme could be a model for other water utilities. The time is now ripe for increased participation of the private sector in other areas of DWASA activities, including the operation and maintenance of water supply and sewerage services.

Establishing a Regulatory Framework

At present there is no specific body to regulate the activities of DWASA, but the government is in the process of establishing a regulatory body for water supply. Such a system will: (1) encourage private sector participation and investment; (2) ensure affordable and sustainable service provision; (3) improve the performance and cost effectiveness of sector service providers; and (4) strengthen the overall transparency and accountability of the sector. However, in the absence of the regulatory body the supervisory ministry has set regulations for municipal water supply which are known as Municipal Water Supply Standard By-laws, 1999 (MLGRD&C , Municipal Water Supply By-laws, 1999). Some elements of these regulations are also applicable to DWASA which are described below:

Tariff regulation. Municipalities/water utilities can set their own tariff structure for metered and temporary connections with permission from the government. However, the Dhaka WASA Board can increase the tariff to a maximum of 5% to compensate for an increase in electricity rates. It should be noted that DWASA spends approximately 35% of its total expenditure on electricity alone. For any increase above 5% for any reason other than the one mentioned above, prior permission from the government is mandatory. For faulty meters and non-metered connections rates can be fixed on the basis of previous three months average water bills. Under the regulations the utilities have been allowed to disconnect water lines of the customers for non-payment of tariffs.

Dhaka WASA collects sewerage tax at a prescribed rate which is equivalent to the water charge. For example, DWASA currently charges US$0.08 per m^3 of water. Holdings with sewer connections have to pay US$0.16 per m^3 of water.

Recovery of arrears. All arrears under the law are subject to recovery under the PDR (Public Demands Recovery) Act 1913.

Water and wastewater quality. The drinking water quality standards have been set by the Environmental Conservation Act (ECA) 1997. Briefly, the rules relating to water quality are that no industrial unit or project shall be established or undertaken without prior environmental clearance from the relevant authority. Whoever contravenes any of the provisions of this Act or the rules or fails to perform duties prescribed under any notice or to comply with any order issued under the Act or rules shall be charged and penalized.

Bangladesh has its own sewage discharge quality standard. Details are given in the Environmental Conservation Rules (ECR) 1997 of the Environmental Conservation Act (ECA). Briefly, the rules relative to the sewage discharge quality are: (1) no industrial unit or project shall be established or undertaken without prior clearance from the Department of Environment; (2) whoever contravenes any of the provisions of the Act or the rules or fails to perform duties prescribed under any notice or to comply with any order issued under the Act or rules shall be charged and penalized.

For the utilities operating wastewater treatment plants it is mandated that the effluent quality of the treated water must meet standards set by the Department of Environment.

Service connection. The urban utilities can provide service connections and install meters to any deserving location where a water connection has been provided. The concerned household has to bear the cost of connection and meter installation. The household has to help urban utilities with meter inspection and meter reading.

Water utilities can provide, on payment by the household, a temporary connection for a period not exceeding one month.

Water/sanitation services provision. Water supplied by urban utilities cannot be diverted for any use other than for domestic purposes. A water connection will not be allowed with a drainage facility of the consumer. Where there is provision of a water supply, no service connection shall be allowed from other sources. The average intermittent supply will be about eight hours a day. Utilities shall not be held responsible for any disruption of the water supply due to natural disasters but shall take necessary measures to restore the system in such cases of *force majeures.*

Households within reach of 100 ft of a sewer line shall have to take a connection and pay taxes or shall have to pay taxes even if no connection is taken. Willful blockage of sewer/drainage lines is prohibited. Construction of drains or diversion of existing drains without permission of the authority is prohibited. Unauthorized connection of house drains to the public drain is prohibited. Unauthorized connection between drainage and sewer lines is not permitted.

Non-revenue water. Non-revenue water continues to remain high at around 54% (DWASA, MIR, December 2003). From 1999–2001 NRW showed a downward trend. In 1999 the UFW was 47% and in 2001 it fell to 40% (Andrews & Yniguez, 2003). Since 2002 the trend has reversed and climbed back to the pre-1996 position.

A study conducted in 1996–97 indicated that about 20% of water was lost due to leaking pipes and joints (GKW Consultants, 1997) and the rest were administrative losses. Major components of administrative loss include non-metered connections, no billing, underbilling, unauthorized connections, pilferage etc.

The principal reason for a reduction of UFW has been the contracting out of the two zones to the employee's cooperative of DWASA which is being managed by the trade unions on behalf of the cooperative together with an aggressive programme for billing and collection in other zones of DWASA.

A comparison of non-revenue water in 18 cities of Asia indicated that Manila topped the list with 62% followed by Colombo, Delhi and Jakarta with 55%, 53% and 51% respectively. The best examples are Osaka (7%), Shanghai (17%), Chengdu (18%), Seoul (25%) and Hong Kong (25%) (Asian Development Bank, 2003). In order to improve the financial position of the utility it is essential that the NRW is significantly reduced to a level of around 30% by eliminating all the probable causes mentioned previously, including the repair of leaking pipes and joints on a priority basis.

The history of non-revenue water indicates that the highest NRW of 75% was recorded in 1980.There was then a steady reduction and the lowest value of 40% was recorded in 2001 (ADB, *Water in Asian Cities*, 2003). From 2002 it started to increase again and in 2003 it crossed the 50% mark (DWASA, MIR, December 2003) and the increasing trend is expected to continue in the short term.

Inadequate capital investment and O & M costs. Comparative data on annual capital expenditure and O & M costs for nine selected Asian cities are shown in Table 2. It can be seen that in the capital expenditure and O & M costs Dhaka ranks fourth in both the categories. Although there are no set rules as to what are appropriate capital investments and O & M costs for a particular utility, it is absolutely clear that the better performing utilities have much higher capital investments as well as O & M costs.

Data presented here are for 2001. In that year the World Bank financed the Fourth Dhaka Water Supply Project which was under implementation and a significant amount of investment was made in the project. The project was completed in December 2002. Since then capital expenditure has significantly declined and there appear to be no major investments available in the near future. If this situation continues for a few years water and wastewater management in the city will decline further from the present unsatisfactory conditions.

Conclusions

The water and wastewater management in the city has not been able to keep pace with the expansion of the city. An over-dependent groundwater based water supply system has nearly exhausted its groundwater resources, threatening an environmental unbalance. Dhaka is the only place in the country where the groundwater is being mined by about 2 m

Table 2. Comparative annual capital and O&M cost for 9 selected countries in Asia

Name of the country	Annual capital expenditure US$/person	Annual O&M cost (US$/connection)
Calombo	1.13	60.33
Delhi	7.78	64.25
Dhaka	2.51	69.94
Ho Chi Minhhcity	4.51	119.22
Hongkong	40.70	326.80
Jakarta	2.95	88.63
Karachi	0.77	30.27
Kualalampur	21.12	301.68
Manila	1.00	100.03

Source: ADB (2003).

every year. A comprehensive conjunctive water management plan, with surface, ground and rainwater has to be implemented to build a sustainable long-term water supply system for the city. The surface water bodies around the city are so polluted, especially during the dry season (January to May,) that it becomes nearly impossible to purify the water for human consumption by conventional water treatment methods. It is therefore essential to take immediate steps to clean these water bodies and revive them as raw water sources for the city water supply. But this is not going to be easy. It is going to require very strong political commitment and large investments. As interim measures raw water can be transported long distances from rivers 45 to 60 km from the city. This will add to the cost of water and particularly hit the lower-income group. A conservative estimate indicated that to meet the water requirements of the city up to the year 2010 approximately US$50 million will have to be invested every year. Neither of these appears to have a place in the government agenda at present.

The status of wastewater management is even worse and is characterized by low coverage and an inadequate and non-functional infrastructure. The master plan prepared for wastewater management calls for an investment of US$40 million per year for the next 15 years. It is unlikely that either the government or DWASA will be able to raise and invest the required funds. It is therefore, essential that the private sector be invited to invest in the construction, operation and maintenance of the infrastructure in the water supply and wastewater management sector.

Unless appropriate steps are taken immediately it may not be possible to attain the millennium development goals of halving the population without access to water supply and sanitation by the year 2015.

References

Andrews, C. T. & Yniguez, C. E. (2003) *Water in Asian Cities: Utilities Performance and Civil Society Views* (Manila, Philippines: Asian Development Bank).

Asian Development Bank (2003) *Water in Asian Cities* (Manila, Philippines: Asian Development Bank).

BUET (2000) *Updating of Existing Groundwater and Land Subsidence Model* (Dhaka: Dhaka Water Supply and Sewerage Authority).

Burger, L. (1998) *Study in Improved Sanitation Service in South Dhaka* (Dhaka: Dhaka Water Supply and Sewerage Authority).

Dhaka WASA (2003) *Management Information Report.*

Dhaka WASA (2000) *Annual Report. 1999–2000.*

GKW Consultants (1997) *Immediate Action Program for Leak Detection* (Dhaka: Dhaka Water Supply and Sewerage Authority).

GOB (1998) National policy for water supply and sanitation. Local government division (Dhaka: Bangladesh Ministry of Local Government, Rural Development and Co-operatives).

JICA (1998) *The Study on the Sewerage System in North Dhaka* (Dhaka: Dhaka Water Supply and Sewerage Authority).

McIntosh, A. & Yniguez, C. E. (1997) *Second Water Utilities Dhaka Book* (Manila, Philippines: Asian Development Bank).

Sir McDonald Partners (1991) *Dhaka region Groundwater and Subsidence Study* (Dhaka: Dhaka Water Supply and Sewerage Authority).

WSP International (1998) *Water Resources and Quality Management Study* (Dhaka: Dhaka Water Supply and Sewerage Authority).

Gold, Scorched Earth and Water: The Hydropolitics of Johannesburg

ANTHONY TURTON, CRAIG SCHULTZ, HANNES BUCKLE,
MAPULE KGOMONGOE, TINYIKO MALUNGANI & MIKAEL DRACKNER

Introduction

The City of Johannesburg in South Africa is an interesting story from a water resource management perspective for a number of reasons. Its existence in the first place is due to a geological factor in which gold-bearing reef was exposed on the surface through a series of ancient biological processes and tectonic events. This means that many of the logical factors that account for the existence of other major cities are simply absent in the case of Johannesburg. Known alternatively as *eGoli* or *Gauteng* (both meaning 'the Place of Gold' in *isiZulu* and *seTswana* respectively), the city of Johannesburg is one of the few great cities of the world that is not located on a river, a lake or a seashore. In fact Johannesburg straddles a major watershed, known as the Witwatersrand (translated literally as 'Ridge of White Waters'), which divides the continent of Africa into rivers that flow into the Indian Ocean to the east, and rivers that flow into the Atlantic Ocean in the west. Located in the headwaters of two major international river basins, the Orange river and the Limpopo river, water supply challenges and water quality issues are but two of the major obstacles that confront the staff of Rand Water, the institution that is responsible for supplying the water that sustains, what is in effect, the economic engine of Africa. The significance of the Witwatersrand Ridge as a life support system relates to the large number of hominid fossils that occur, with some 40% of the world's known hominid fossil deposits being associated with this geological feature. So while Johannesburg is a modern African city, it is also the Cradle of Humankind[1]. Today the Witwatersrand metropolitan area with a population of around 11 million people, largely dominated by Johannesburg and the many satellite towns and cities spawned by the gold mining activities of the last century, is said to be one of the largest concentrations of humans that has developed away from a sustainable water resource. Rand Water, the statutory body responsible for providing potable water to this sprawling urban conurbation, is one of the largest bulk water suppliers in the world today. Therefore, the story of water in Johannesburg is the story of Rand Water, which in turn is about sustaining a major urban conurbation that is far greater than the actual limits of this one city alone.

Physical Aspects

The City of Johannesburg is found at an altitude of some 6000 feet (1800 m) above sea level (Figure 1). It is located in an area that was originally a rolling grassland called the Highveld, which was devoid of any trees except for those found along the riparian zones of the many small rivers that are found there. The grassland is largely the result of the low and highly variable rainfall patterns, with a mean annual precipitation (MAP) of 600 mm. The Highveld covers a semi-arid area with evaporative losses of 1600 mm. The low MAP, combined with the high evaporative potential and variability in precipitation patterns, has resulted in the fact that rivers in South Africa have amongst the lowest conversion of mean annual precipitation to mean annual runoff (MAR) in the world (O'Keeffe *et al.*, 1992, p. 281). In fact, the total average runoff (that portion of rainfall that is not lost to evaporation that eventually finds its way into rivers) is only some 10% of total annual rainfall (Rabie & Day, 1992, p. 647) in South Africa as a whole. Of this relatively small runoff that eventually becomes streamflow, a mere 60% (Rabie & Day, 1992, p. 647) to 62% (O'Keeffe *et al.*, 1992, p. 278) can be economically exploited due in large to the extreme variability of the rainfall events. This hydrological reality has provided the stimulus for the construction of dams in an attempt to retain as much streamflow as possible (Turton & Earle, 2005), to the extent that South Africa is listed in the top 10 countries in the world by virtue of the number of large dams that have been constructed over time (WCD, 2000, p. 373; Turton, 2003b).

Cutting across the Highveld area is a geological feature that is the result of a series of tectonic events. This geological feature consists of recurring sedimentary cycles which have passed in phases of geological time, from deep sea marine to continental conditions (Werdmüller, 1986, p. 37). The gold-bearing reef within this geological feature has been described as being 'typical lagoonal deposits' that have been the subject of tectonic movements in which subsidence, uplift and subsidence has occurred, along with substantial weathering and erosion (Werdmüller, 1986, p. 37). More specifically, what is

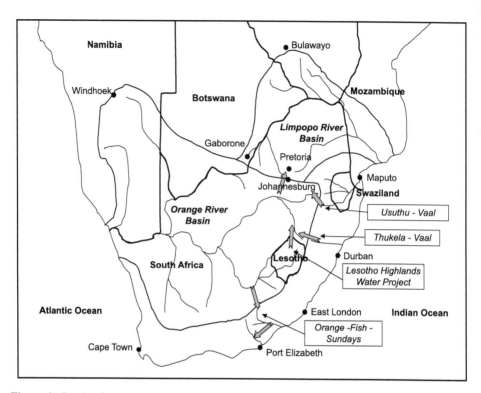

Figure 1. South Africa showing the Orange and Limpopo River Basins, with Johannesburg on the watershed and some of the strategic Inter-Basin Transfers

known as the Main Reef Leader is attributed to shore line pebble deposits in which gold and other heavy metals became concentrated during the gradual but relentless advance of an encroaching sea front (Werdmüller, 1986, p. 37). The inland sea that created these deposits dates back to periods of geological time between 3100 million years and 1900 million years ago (Werdmüller, 1986, p. 44). The entrapment of gold has been helped by the existence of primitive plants and algal-like material, which are preserved today as a thin band of carbon (Chapman *et al.*, 1986, p. 181) that is known as the Carbon Leader.

The Witwatersrand Ridge is important for five key reasons. First, the tectonic event that created it also created the major continental watershed, with rivers arising to the east flowing into the Indian Ocean through the Limpopo river basin, and rivers arising to the west flowing into the southern Atlantic Ocean through the Orange river basin. Both of these are strategically important to South Africa because of the high level of economic activity that they sustain.

Second, the tectonic event exposed gold-bearing reef to the surface, thereby making it relatively easy to discover once it was settled by farmers in the 1800s. The discovery of the Witwatersrand goldfields has probably exerted a greater influence on the evolution of South African history than any other single event (Cunningham, 1987, p. 7).

Third, the name 'Witwatersrand' gives an indication of the early pristine condition of the Highveld area. Translated literally as 'Ridge of White Waters', the name arose because of the myriad of small springs that rose along the geological feature, many of which cascaded across the different elements of the geological complex before starting their long journey as rivers down to the sea many thousands of kilometres away. The Witwatersrand is thus an important headwater for two major river basins, so pollution occurring there impacts a wide range of stakeholders downstream.

Fourth, the continental watershed created by the Witwatersrand, is part of a regional Hydropolitical Complex consisting of the four most economically developed states in the Southern African Development Community (SADC) region (South Africa, Namibia, Botswana and Zimbabwe), all of which have water scarcity constraints to their future economic growth potential (Turton, 2003a, 2003b, 2004; Turton & Earle, forthcoming; Turton *et al.*, 2004a, 2004b). Central to this complex are the Orange and Limpopo river basins, which are shared by these four riparian states, making these two basins pivotal in the sense that their strategic significance goes way beyond simply sustaining the socio-economic activities in the city of Johannesburg[2].

Finally, the Witwatersrand Ridge offered a safe haven for early hominid development. Providing high ground with a panoramic vista over the plains of Africa, early hominids could hunt the migrating herds of antelope. Significantly, the Highveld grassland biome has been described as sustaining the highest animal biomass of any similar ecosystems in pre-colonial Southern Africa (Mitchell, 2002, pp. 18–19). The underlying dolomite geology provided caves in which shelter could be found, the most notable of which is Sterkfontein Cave (translated literally as 'Strong Fountain Cave') in the Krugersdorp area of the West Rand. It was consequently along this watershed that human development was encouraged in pre-historic times, with fossil discoveries dating back to some of the oldest known hominid branches on the African continent. This is a World Heritage Site[3] today with the Cradle of Humankind as a component.

The significance of the combination of these key factors is that Johannesburg exists entirely as the result of the discovery of gold, is located on a watershed at high altitude, and lacks any of the other fundamental factors that are usually associated with the development of major cities elsewhere in the world. Johannesburg, in this sense, is very unique indeed, being one of the few cities in the world that is not located on a river, a lake or a harbour. However, more significantly Johannesburg straddles a major watershed that has strategic ramifications for four of the most economically developed states in the SADC region, all of which have reached a water-related limitation to their future economic growth potential. This combination of factors is the underlying driver of the many challenges that arise in the quest to manage the water resources that sustain the economic growth and development in and around the city.

Driver of Growth

The geology associated with the Witwatersrand Ridge, exposed gold-bearing reef to the surface, thereby enabling gold to be discovered. It is here that the original driver of growth can be traced. In 1885 two wanderers, George Harrison and George Walker, arrived on the farm Wilgespruit (translated as 'Willow Stream') on the northern slopes of the Witwatersrand Ridge. Early in February 1886, Walker left Wilgespruit to join Harrison, who was building a house on the farm Langlaagte. En route to Langlaagte Harrison stumbled on an outcrop of rock, and after examining it, decided to crush and pan a piece of it. Having been a gold digger in Australia, he had the necessary skills to recognize the reef, and was surprised when he saw a tail of gold in the pan. He had inadvertently stumbled onto a weathered outcrop of what was to become known as the Main Reef (Werdmuller,

1986, p. 7). This was the start of the Gold Rush[4] that saw Johannesburg grow virtually overnight.

Similar discoveries soon followed with small mining towns springing up all along the Witwatersrand Ridge. The mining camps were generally sited according to the availability of water (Appelgryn, 1984, p. 14). The East Rand saw a cluster of mines starting to coalesce around what were to become future satellites of Johannesburg. Located at places like Springs (referring to the original freshwater springs that existed there), Brakpan (translated as 'Brackish Pond') and Benoni. In similar vein the West Rand spawned a plethora of mining towns in places like Roodepoort (a 'poort' is a natural ravine through a mountain caused by a river and often associated with a geological fault), Krugersdorp[5] (named after Paul Kruger, the President of the Zuid Afrikaanse Republiek when gold was discovered) and Randfontein (translated as 'Ridge Fountain'). The situation that arose shortly after the original discovery of gold was a typical gold rush bonanza, with the rapid influx of foreign diggers, capitalists and would-be prospectors. They settled in a series of tented and corrugated iron towns around each of these main discoveries, dotted all along the Witwatersrand Ridge. Over time these small settlements merged into one bigger conurbation, and on 4 October 1886 the City of Johannesburg was officially founded[6], subsequently accounting for some 40% of the global production of gold.

Politically this caused significant tensions. The Witwatersrand Ridge was located in an independent Boer Republic known at that time as the Zuid Afrikaanse Republiek (ZAR) with Paul Kruger as the President. The ZAR was one of two Boer Republics, which had arisen as the result of the Great Trek away from the British colonies of the Cape and Natal. Happy to settle the land and farm relatively free of foreign intervention, the Boer Republics were given a rude awakening when literally thousands of *Uitlanders* (translated as 'foreigners') descended upon them, all seeking wealth from the gold bonanza. This changed the politics almost overnight. Once it had been established that the gold discoveries were substantial, British interest was soon raised. The hidden wealth of the ZAR was to be coveted by the Crown[7]. A series of political interactions occurred, the most notable of which was the Jamieson Raid[8], in which the British raised a band of mercenaries who attempted to unseat Paul Kruger by force. In a series of skirmishes that saw the raid fail, Paul Kruger retained political control for a while. This was only a temporary respite however, as the subsequent Anglo-Boer War[9] eventually saw the ZAR being incorporated by military force into what became the Union of South Africa. Consisting of the two former British colonies of the Cape and Natal, along with the two vanquished Boer Republics of the Orange Free State and ZAR (also referred to as the Transvaal), the Union of South Africa was founded largely on the mining wealth arising from the original discovery of gold at Langlaagte, and diamonds[10] in the kimberlite pipes around the modern city of Kimberly.

Recent History

At the time of the establishment of the Union of South Africa in 1910, the potentially promising gold-based economy was largely shattered. The Boers had been vanquished as the result of the scorched earth policy of the British, which saw large scale ransacking of the farms and the first recorded use of concentration camps[11] for women, children and other non-combatants on the African continent. Union was therefore a bitter affair for many Boers, who had become landless peasants without any modern skills, largely because of the perceived greed of other non-African citizens. However, the newly established mining houses created a demand for labour so there was a general influx of Afrikaners[12], many of whom who had been systematically dispossessed during the Anglo-Boer War. There was a demand for four distinct categories of worker on the rapidly developing mines—unskilled labourers, semi-skilled miners, specialists such as geologists and engineers, and entrepreneurs. These four categories had a specific impact on the development and growth of Johannesburg, and indeed the social fabric of South Africa as a whole.

The entrepreneurial class were all *Uitlanders*, mostly from Britain. This was so simply because the Boer Republics had not yet mobilized any financial capital or entrepreneurial skills needed to fund and manage the deep mining operations. The entrepreneurial class settled in a wealthy new suburb that was known as Doornfontein (translated as 'Thorn Fountain').

The skilled specialists such as geologists and engineers were also originally sourced from outside South Africa, coming mostly from the Universities of Britain (but also from other European countries). They generally spoke English.

The semi-skilled miners came mostly from the local Boer population. They spoke Afrikaans[13] and other indigenous African languages such as *isiZulu* and *seTswana*, which they had acquired in their earlier lives on the rural farms. As such they became a vital link in the management chain between the entrepreneurial class and skilled specialists that directed the mining operations and the unskilled labourers. They generally settled on the mine properties.

The unskilled labourers were all native Africans, speaking a variety of indigenous languages. They were used to wrestle the gold from the deep reefs which they followed relentlessly over time, eventually going thousands of metres into the earth in what were destined to become some of the deepest mines[14] in the world, often paying with their lives. These unskilled labourers were housed in hostels on the mine properties.

It was the need for labour on the mines that became one of the major drivers of the subsequent social fabric of South Africa that had two distinct components to it. Central to this was the notion of migrant labour, with men going to Johannesburg to earn wages to be remitted back to the rural areas all over the continent of Africa. This left the women back home, tilling the fields and raising a family in which men were hardly seen and thus increasingly irrelevant. The men on the other hand, away from home for 11 months of the year, lived in all-male hostels. The migrant labour system became deeply entrenched in the South African economy and society, with a range of debilitating features that were to become evident many decades later. These maladies manifest themselves as the social pathologies of the breakdown of the family unit, alcohol and substance abuse and prostitution. It was to these fertile soils of social disruption that the destructive seeds of HIV/AIDS were to be sown nearly a century later, by which time the social structure of South Africa had been so severely damaged that the pandemic could take root and spread. The first component of the South African social fabric thus consists of migrant labour, which has become so deeply entrenched today that it is impossible to understand Johannesburg without reference to this key factor.

The second component is the history of Apartheid, which also has its roots in the discovery of gold. The stratification of society that led eventually to the formal policies of Apartheid being launched in 1948 can be traced back to the social composition of the mining houses in the immediate post-Anglo-Boer War era. The landless Afrikaner white minority who had lost nearly everything because of the British scorched earth policy, found a new niche on the mines. Eager to protect their positions, and not yet ready to occupy the specialist categories through lack of appropriate tertiary education, and the entrepreneurial categories through the lack of capital, they began introducing legislation that came to be known as Job Reservation. Becoming a cornerstone of the subsequent Apartheid policies, Job Reservation saw every task being classified in terms of racial composition. Unskilled jobs were left to the Black majority, while Whites reserved jobs involving the management of mining machinery and explosives to themselves. In terms of this approach, Blacks were not allowed to handle explosives, or to handle machinery such as hoists. This became a cornerstone of Apartheid half a century later.

It was against this social backdrop that Johannesburg grew. As this occurred, the dust associated with the Highveld climate became a problem, especially when whipped off the growing mine dumps that were starting to dominate the landscape, so trees were planted (Figure 2). The first public park was established in 1887 in what is still known today as Joubert Park (Davie, 2003). As Doornfontein was outgrown by the magnates, new suburbs were developed in Victorian fashion, laid out with large formal gardens and complex public parks. The new posh suburbs to which the so-called Randlords migrated reflect this fact in their names—Park Town, Parkview and Forest Town (Davie, 2003). Trees were planted in vast numbers, with demand coming from two distinct sources. The first was the need to settle the dust by creating parks for recreational purposes. The second was the need for mine timbers to support the shafts and stopes that were being driven ever deeper into the Reef, following the Carbon Leader like an indelible pencil mark permanently etched on the rock face of geological time. This saw vast quantities of Eucalyptus trees from Australia being planted, which later became a major environmental problem as they choked the small rivers arising along the Witwatersrand Ridge desiccating the landscape.

Population growth became a major factor in the water management equation. From a sparsely populated rural setting before the discovery of gold in 1886, the population had grown to 100,000 a decade later, and to 150,000 by 1901. This had mushroomed to 420,700 by 1910, and to over 10 million by the 2000.[15]

Role of Water in Sustaining the Greater Johannesburg Conurbation

Shortly after gold was first discovered, water for the small mining town was drawn from a spring called Fordsburgspruit (translated as 'Ford's Mountain Spring'), and a second fountain near the present Johannesburg General Hospital called Natalspruit. Prior to 1896 another source of supply was developed on the farm Weltevreden (translated as 'Very Satisfied') north of the mining operations in Roodepoort. When this failed to meet the demands, development turned to dolomite groundwater found on the farm Zuurbekom, with a pump station first erected there in 1899 that is still in operation today[16] providing high quality water that is not in need of any treatment before being mixed with other sources. The growing demand for water was driven by the accelerating need to process crushed ore[17], with about 2000 litres (2 tons of water) being needed to mill one ton of gold-bearing reef. In the early 1890s a new process of extraction was introduced. Called the McArthur Forester process, it used cyanide to coalesce the fine gold particles, thereby increasing the yield of recoverable gold from the milling process, but also resulting in a potential pollution hazard that still exists today.

The voracious demand for water saw the first major private concession being granted by the ZAR Government in 1887. Known as the Sievewright Concession, one of the

Figure 2. The modern Johannesburg skyline showing the large number of exotic trees. This is the largest man-made forest in the world registering on satellite images as a tropical rainforest, posing specific challenges to water supply engineers today. *Source:* Rand Water.

stipulations was a limitation on the price that could be charged for water. The source of water used for this was the Doornfontein springs and the concession gave rise to the Johannesburg Waterworks and Exploration Company. This company was purchased in 1889 by mining magnate Barney Barnato. By 1893 the demand for water was around 5.86 Ml/d, outstripping the supply capacity of the existing companies. In 1893 the Braamfontein[18] Water Company, supplying water to the newly-established upmarket suburb of Parktown, and the Vierfontein[19] Syndicate, supplying water of different quality to industrial and human users, were established. In 1889 the Vierfontein Gold Mining Company constructed the first pumping station in the Klip river valley, receiving its water from the Olifantsvlei[20] Farm near present day Turffontein. This grew over time into what became known as the Klip river scheme. This led naturally to the Vaal river, some 60 km away from the goldfields, being considered as a reliable strategic supply of water for what was by then being called 'The Reef' that was centred on Johannesburg. A concession was awarded to two engineers in 1889 to deliver water from the Vaal river. One of these engineers, F.C. Eloff, was the son-in-law of President Paul Kruger, so the first allegations of nepotism were raised in the water supply story of Johannesburg. These died out when in 1892 the Barnato brothers' Johannesburg Waterworks and Exploration Company gained control over this scheme. This was the foundation of the linkage between water supply and economic wealth that is the cornerstone of the so-called Pipelines of Power thesis[21] (Turton, 2000) on which an explanation of the hydropolitical dimension of Apartheid is based. By the end of the Anglo-Boer War in 1902 there were three companies responsible for water supply in the Witwatersrand–Johannesburg area—the Johannesburg Waterworks Estate and Exploration Company, the Braamfontein Company and the Vierfontein Syndicate—sourcing water from the dolomites around Zuurbekom, the springs at Doornfontein (near modern day Ellis Park stadium), a well an spring at Natalspruit in central Johannesburg, a spring in Berea, a spring in Parktown and the Klip river valley pumping station in the south. It was becoming apparent that the availability of clean water at a high assurance of supply could become a limitation to the economic growth of Johannesburg, so additional sources were considered for strategic reasons.

Severe droughts occurred in the 1890s triggering a series of water crises. Water carts came into operation, selling exorbitantly priced water that was largely unaffordable to most of the non-industrial users. This coincided with the introduction of the McArthur Forester process of gold extraction, which saw the first cyanide contamination of rivers. The Klip river was polluted in 1894 resulting in the death of livestock drinking the water. A Commission of Enquiry was established, which reported that water in the Doornfontein Valley was also polluted by mine effluent. This was the birth of a substantial water resource management problem that persists to this day, a fact made worse because the rivers being polluted are the very headwaters of major river basins that cross international borders.

In 1901 the ZAR Military Government appointed the Witwatersrand Water Supply Commission. At the end of the Anglo-Boer War in 1902 the British had gained control over Johannesburg and they realized that stable local government had to be restored if future economic growth was to be realized. The Royal Engineers took initial control in investigating water supply and sanitation services. The Witwatersrand Water Supply Commission reported in 1902, making the recommendation that an institution to be called The Rand Water Board[22] should be established and given the responsibility of developing a secure water supply system. The Rand Water Board included members of the Johannesburg Town Council, the Chamber of Mines and representatives of other local authorities along the Witwatersrand Ridge. The formal establishment of the Rand Water Board[23] occurred through Incorporation Ordinance No. 32 in May 1903 with Lt. Gen. George Fowke as the Chairman. Representatives on the Board consisted of three nominated members from the Johannesburg Town Council, five members from the Chamber of Mines with other members coming from Boksburg and Germiston on the East Rand, and Krugersdorp and Roodepoort on the West Rand. This laid the formal foundation for stable bulk water supply to all of the key towns along the Witwatersrand Ridge that were to eventually grow into what can be described as the Greater Johannesburg Conurbation with no clear geographic distinction between the different components. Official talks[24] between Pretoria City Council and the Rand Water Board in 1928 paved the way for the incorporation of Pretoria into the water supply network being established

Projections made by the Transvaal Chamber of Mines for the years 1915–1920 indicated that mining needs for water would range between 48 Ml/d–70 Ml/d. Domestic demand was projected to be in the order of 16 Ml/d–20 Ml/d. This meant that a combined demand of 90 Ml/d had to be planned for by Rand Water Board by 1920. Severe water shortages in 1913 focused attention on the need to secure the water supply if the full economic potential of the gold discovery was to be realized. This acted as a trigger event that changed the water management paradigm away from sourcing water from local springs and streams, to the Vaal river some 60 km away. In 1914 the Rand Water Board adopted the Vaal River Development Scheme based on a phased approach. The first phase consisted of the construction of the Barrage along with purification works and pumping stations in Vereeniging and a main pipeline to the Witwatersrand. As this scheme was being implemented the Great War broke out in Europe, placing a restriction on the possibility of raising funds. The Board sent a deputation to Europe and Egypt to investigate possible solutions. Investigations also included water supply planning from the Mississippi river in the USA. Construction of the Barrage was commenced in 1916 and completed in 1923. The security of water supply that this established meant for the first time that the full economic growth potential of the goldfields could be realized with confidence. Additional confidence arose when in 1934 the Vaal River Development Scheme Act was passed by Parliament, paving the way for the construction of the Vaal Dam, which was built during the Great Depression[25]. The yield of the Vaal Dam[26] was increased in 1955/6 to result in a capacity of 2330 million m^3, or more than twice its original volume.

In 1940 a critical water shortage in Pretoria saw the reopening of negotiations with the Rand Water Board. After protracted negotiations Pretoria was represented on the Board and the water supply area was officially increased to provide the foundation for industrial development in that city. This gave significant impetus to the development of an urban conurbation between Johannesburg and Pretoria, resulting in the fact that the two cities are for all intents and purposes one place today, despite the fact that they are managed by two different local authorities. Running concurrent with these negotiations was the need of Vereeniging to be included in the supply network. Both Vereeniging and Pretoria were officially incorporated into the Rand Water Board supply area simultaneously in 1944. This gave rise to the so-called PWV Triangle[27] in which the largest concentration of heavy industry and economic activity on the continent of Africa is based.

In 1960, roughly coinciding with the independence of South Africa and the ending of British colonial rule, the Rand Water Board increased its water supply area through the approval of the Eastern Transvaal Water Supply Scheme. This enabled the strategic oil-from-coal development to take place with the subsequent establishment of the SASOL plant at Secunda on the Far East Rand. This meant that the available supply was stretched to breaking point, so alternative strategic sources of water were investigated, leading to the Thukela–Vaal Augmentation Scheme[28] being launched in 1974. This was the first inter-basin transfer from a river basin outside of the normal catchment of the Vaal system. Running almost concurrently with the feasibility of the Thukela river as an alternative source of supply, investigations were launched into the possibility of taking water from the highlands of Lesotho. This gave rise ultimately to the Lesotho Highlands Water Project (LHWP) with the first water flowing into the Vaal Dam via the Ash river outfall on 8 January 1998. Details of current strategic supply via inter-basin transfers are shown in Figure 3, while the hydropolitical history timeline for Johannesburg is shown graphically in Figure 4.

Role of Third Parties

The Johannesburg case study shows that third parties have not played a significant role in water resource management.

Future Prospects

The Johannesburg case study suggests the existence of three strategic issues of contemporary relevance. These are issues of supply, issues of quality and issues arising from the fact that Rand water impacts on two pivotal basins in the Southern African Hydropolitical Complex. Each of these is discussed separately below.

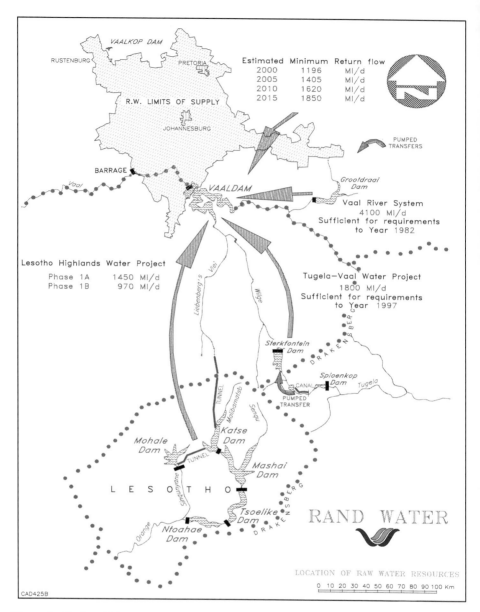

Figure 3. The Rand Water Supply Area showing some of the major inter-basin transfers needed to sustain the Greater Johannesburg Conurbation. *Source*: Reproduced with permission of Rand Water.

Issue of Supply

The growth in real consumption and supply by Rand Water since 1903 is represented by the lower line in Figure 5. The upper line depicts the different supply schemes that were developed and built to be able to cope with the demand increase. It can be seen that from the 1980s onward, that engineering solutions, mostly in the form of inter-basin transfers from the Thukela river and the Lesotho Highlands Water Project, kept ahead of demand. This means

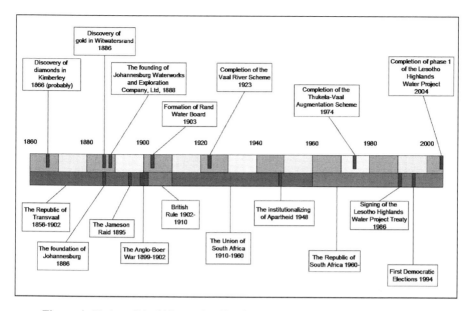

Figure 4. Hydropolitical history timeline for Johannesburg showing major events

that there is some overcapacity available in the supply system of Rand Water at the time of writing. The main cause for overcapacity is related to two distinct aspects. First, there were substantial droughts in the early 1980s, and again a decade later in the early 1990s, which resulted in the implementation of Water Conservation (WC) and Water Demand Management (WDM) strategies. These curbed the growth in demand. Second, the economy did not grow

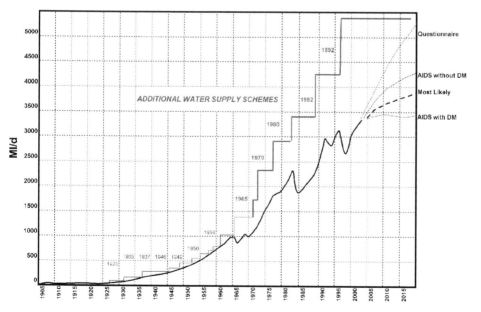

Figure 5. Water Demand and Supply curves for Rand Water over the lifespan of the water supply utility. *Source*: Data courtesy of Rand Water Engineering.

at the anticipated pace, possibly as a result of the uncertainty arising from the collapse of Apartheid and the transition to democracy in 1994, and also possibly as the result of the impact of HIV/AIDS on the economy as a whole. The exact nature of this has not yet been determined so this aspect remains the subject of some speculation today.

The future projections for water demand have been broken down into four possible scenarios by Rand Water. Scenario 1 is the most optimistic, based on strong economic growth and no HIV/AIDS impact. Scenario 2 factors in the impact of HIV/AIDS in so far as the pandemic has been quantified, but assumes no additional WDM measures will be implemented. Scenario 3 is also based on the best estimate of HIV/AIDS impact, combined with additional WDM strategies to curb consumption. Scenario 4 is the most likely trajectory based on the best available current knowledge about HIV/AIDS. This shows a growth in demand but at a reducing rate, suggesting that aggressive water supply augmentation is not likely in the mid-term future. This has already had an impact on the Lesotho Highlands Water Project, where Phase 2[29] has been placed on hold. This has reduced the income stream to Lesotho as royalty payments are based on water volumes delivered. The feasibility study[30] of using the existing infrastructure in the Thukela–Vaal Pumped Storage Scheme is currently underway. This provides a least-cost option should future demand in the Rand Water supply area need to be augmented. Should this option be followed, two new dams will be constructed on the Thukela river, Jana and Mielietun Dams, feeding into the existing Pumped Storage Scheme. This decision has other strategic considerations beyond mere water supply, as the Pumped Storage Scheme is an integral component of the national electricity grid, using surplus energy in off-peak periods to pump water across the Drakensburg Mountains, to recover some of that energy again when water is released into the Upper Vaal in order to meet peak energy demands. The additional volume of water will thus provide greater flexibility in managing the peaks in energy demand at the national level.

Central to Rand Water's planning is the need to understand the impact of HIV/AIDS. The Chief Engineer (Development) of Rand Water generated a document titled 'Rand Water's Integrated Least Cost Planning Model' in May 2000. In this document the problematique of HIV/AIDS is analyzed with respect to the provision of infrastructure (Figure 6). From this assessment, two key strategic issues arise:

- At the point of the peak demand prior to population reversal, the infrastructure could be under-utilized in the long term.
- At the point of minimum demand after population reversal, attempts to 'ride through' the peak demand by imposing water restrictions would offer the maximum utilization of infrastructure.

After careful consideration of these critical issues, the report concludes that both options have high-risk profiles for the Gauteng area. The first option raises the risk of financial crisis arising from excessive investment in infrastructure. The second option raises the risk of non-supply[31] at crucial periods of time. The report also suggests that regions with strong economic activity and high employment rates may not experience significant population reversal. This introduces some uncertainty into the overall water management regime for the Greater Johannesburg Conurbation.

One of the ramifications of this uncertainty suggests that even if infrastructure is provided to meet a growth in demand of 2.76% per annum, the infrastructure may be provided four to nine years earlier than necessary. Water restrictions may have to be used

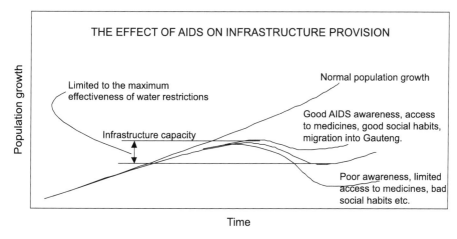

Figure 6. The effect of HIV/AIDS on infrastructure planning as envisaged by Rand Water

to 'ride through' the point of peak population. This is a complex decision as it would have to be based on the as yet unknown economic impact that HIV/AIDS has in the region, the financial impact that new infrastructure would have on a declining income-base and an understanding of which user groups would be mostly affected by HIV/AIDS. The reality is that there is currently insufficient information at a high enough level of confidence to make any strategic decisions about infrastructure provision based on the impact of HIV/AIDS in the Greater Johannesburg Conurbation.

Within this set of scenarios it is difficult to cater for WDM in the forecast. In the Greater Johannesburg Conurbation, the unaccounted-for-water figure is around 50%. During the anti-apartheid struggle, non-payment for services was used as a tool by disenfranchised township inhabitants to overthrow a repressive regime. With the birth of democracy in 1994, an era of unprecedented optimism was ushered into the political history of South Africa. However, within the context of water supply there were some unforeseen consequences that are still having a lasting impact on the development of sustainable water supply for the Greater Johannesburg Conurbation. Central to this is the culture of non-payment for services, which was such a central part of the so-called 'struggle'. Several studies done by the Data and Research Manager of Rand Water have shown that there are widely differing reasons given by respondents to non-payment for services, which includes unemployment, insufficient disposable income, deep-seated mistrust for government and the distance to pay points. This survey has also shown that when there is insufficient disposable income in a given household, payment goes towards electricity rather than to water, due partly to the fact that the latter is enshrined in the Constitution and may not summarily be cut off. Other causes include the entertainment factor in the use of electricity. However, this is not the sole cause for non-payment which include: the amount of disposable income being eroded by the use of cellular telephones; the burgeoning of gambling and the introduction of Lotto. The combination of these is a reduced income stream to Rand Water, which in turn has substantial ramifications in terms of future service delivery.

The need for reversing this trend is exacerbated by the massive water losses that occur in the area. Water wastage to the value of US$378 000 occurs daily in the Rand Water area of supply. Water wastage in this sense refers to potable water that runs into the ground from leaking

High Volume   Low Volume
Users   users

Figure 7. Matrix of Water Demand Management target group priorities used by Rand Water

reticulation networks of municipalities before being metered by a given consumer, or that runs out of leaking taps and toilet cisterns on private property after being metered. Recent work by Rand Water has shown that 80% of the wastage occurs on private property.

Rand Water recently completed an investigation into water losses and found in a township of 63 000 families that the mean daily average demand of the area was 3500 kilolitres per hour. Measurement of the demand showed that the Minimum Night Flow, that is the amount of water entering the area between 02:00 and 04:00 in the morning, amounted to 2700 kilolitres per hour. This shows a substantial loss in the reticulation system.

The Rand Water WDM strategy is based on a process of prioritization using the matrix shown in Figure 7. Non-paying users of high volumes of water are tackled first (Quadrant 1), followed by smaller non-paying users (Quadrant 2). The primary management objective of these two target groups is to bring them into the income stream of Rand Water, thereby securing the financial viability of the infrastructure as a whole, while using the price elasticity of water to reduce excessive demand. Next come high volume users that pay for their water (Quadrant 3), but herein lies a dilemma, because while these consumers are the ones that use water inefficiently, they also provide the income stream to cross-subsidize the losses in other parts of the system. For this reason Quadrant 3 will only be tackled under drought conditions, and then usually with retrofitting as an objective. Finally, smaller users that pay for their water are the last to be targeted, mostly with retrofitting and increased efficiency as an objective.

Experience on the ground is showing that WDM is a complex issue to manage and effectively implement. In the immediate post-apartheid era of service delivery, the focus is mainly on improving supply to the people that traditionally did not have access to water and sanitation. This also happens to be the poorest segment of society with the least capacity to pay. It is therefore seen as a political imperative that funds be spent on such service delivery. From purely a political perspective then, it looks good if, at the end of a project for the supply of water and sanitation, a grand opening can be launched where a tap is opened and another sector of the population has access to water and sanitation. The rewards are simply not the same for a WDM project, even if it has put back millions of litres of water into the system which otherwise would have been lost. There is no ribbon to cut in public at the end of the day, and WDM often comes at the price of punitive water bills or reduction in consumption. These are too closely related to hardship to translate into political kudos, so politicians seem to avoid such measures.

Issue of Quality

Water quality issues are becoming increasingly important for a variety of reasons. The first major cluster of issues in this regard relates to the chemistry of groundwater in mines (Ashton[32], personal communication). Given that the majority of the mining operations have occurred along the Witwatersrand Ridge, the water table has been substantially reduced due to mine dewatering activities. Water arising from mining operations tends to come into contact with the residue of the mining operation such as dust, exposed reef and waste rock.[33] Depending on the type of minerals being mined, this water dissolves chemical substances in that residue. When sulphite ores, particularly those containing pyrite (often associated with gold-bearing reef) come into contact with a combination of air and water, they give off sulphur dioxide and produce low pH waters with a high concentration of iron and sulphate. This process is accelerated by specific groups of bacteria that help to transform inert sulphides into acidic solutions. This problem is of a significant magnitude, particularly when mines reach the end of their useful lives and are closed, known as acid rock drainage. In the South African sector of the Limpopo river basin for example, there are over 1000 abandoned mining operations, though most of these are small gravel quarries and clay pits. The metal mines, mostly lead, tin and zinc, are small 1–10 man operations so their impact is negligible, but the larger coal, base metal (copper, lead, tin, zinc, chrome, vanadium) and precious metals (gold, platinum) tend to have an acid rock drainage problem (Ashton, personal communication).

Another different category of water quality problem relates to the effluent drainage from the Rand Water supply area. Associated with this are nitrogen (N) and phosphate (P), with the ratio of the two being a critical parameter (van Ginkel[34], personal communication). The N:P ratio is important for the community structure of phytoplankton, specifically with regards the relative competitive capabilities and biomass of cyanobacteria in epilimnetic phytoplankton communities (Figure 8). Low N:P ratios indicate the presence of cyanobacteria in phytoplankton communities. There is some debate over the exact dynamics of this, particularly with respect to the question as to whether the N:P ratio drives the dominance of cyanobacteria, or whether it is the result of the presence of cyanobacteria in the system. In this regard Wentzel (2001) has shown that cyanobacteria tend to be rare at N:P ratios of greater than 29. Van Ginkel (personal communication) has noted that this may be true for some species including *Microcystis*, but may not be true for all of the cyanobacterial species. Van Ginkel (personal communication) concludes that there is definitely some relationship between the N:P ratio and the development of specific phytoplankton species, but notes that care must be taken in interpreting the ratio without taking cognisance of the fact that blooms are multivariate in origin.

The significance of this is that nitrogen and phosphate tends to become concentrated in dams that receive effluent streams from sewage water treatment plants. For example, Hartebeespoort Dam is the first dam on the Crocodile river, a tributary of the Limpopo river, which receives treated effluent streams from both Johannesburg and Pretoria.

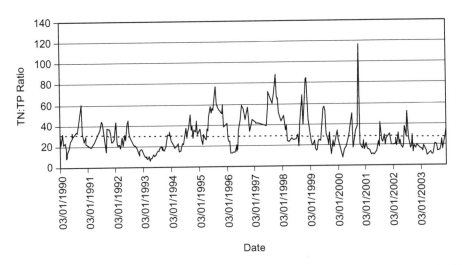

Figure 8. Time series of the N:P ratio for Hartebeespoort Dam. Low N:P ratios are associated with severe cyanobacterial blooms in South Africa

Studies of this dam[35] show that a low N:P ratio is associated with highly eutrophic systems dominated by cyanobacterial blooms. The most dominant bloom-forming cyanobacteria in South Africa is *Microcystis* (van Ginkel, personal communication). Venter (2004) notes that at least 20 tons of phosphate flows into the dam each year, so even if current treatment methods could reduce these levels at point source, they would not cater for the vast amounts of phosphate already trapped in the sediment of Hartebeespoort Dam, which is acting as a source for future releases downstream. This raises the huge technical problem of reducing the phosphate level at the dam from an average of 0.12 mg/l to less than 0.05 mg/l in order to significantly reduce future pollution loading of the headwaters of the Limpopo river basin. The cost of this technology is simply prohibitive for a developing country like South Africa, so it is unlikely to be introduced in the near future. This also illustrates the need for constant technical innovation in highly modified ecosystems.

Issues Arising from the Southern African Hydropolitical Complex

Research work under-way at present suggests the existence of an emerging hydropolitical complex in Southern Africa (Turton, 2003a, 2003b, 2004; Turton *et al.*, 2004a, 2004b). Central to the rationale of this concept is the fact that water scarcity is reaching the point where four of the most developed countries in the SADC region—South Africa, Botswana, Namibia and Zimbabwe—are all reaching a point where future economic growth and hence political stability can be threatened. This makes the issue one of strategic importance. Within the so-called Southern African Hydropolitical Complex there is sufficient evidence to suggest that water scarcity constraints are becoming the trigger for co-operation rather than conflict, and are reaching sufficient magnitude to impact on the international political relations between the respective sovereign states in the region. Two international river basins of the most strategic importance within the Southern African Hydropolitical Complex are the Orange and the Limpopo, which have been called pivotal basins. With Johannesburg being located on the watershed between these two pivotal basins, the water supply imperatives link the two by way of IBTs and return flows. The reasons for the strategic significance of these two river basins are the facts that:

- both support substantial economic activity within each of the riparian states;
- both are shared by the four pivotal states in the Southern African Hydropolitical Complex (South Africa, Botswana, Namibia and Zimbabwe);
- both have reached the point where there is simply no more water left to be allocated to existing economic productive activities;
- both have significant water quality issues impacting on their future ecological sustainability.

Consequently, what happens in the future water resource management approach by Rand Water in sustaining the economical viability of the Greater Johannesburg Conurbation, will automatically cascade downstream and impact on the strategic options of the four pivotal states in the Southern African Hydropolitical Complex. This makes the Orange and Limpopo pivotal basins within the Hydropolitical Complex.

Lessons Learned

The Johannesburg case study provides three valuable lessons learned.

The first major lesson is that complexity tends to breed further complexity. The initial complexity arising from the need to supply water to a city at high altitude straddling a watershed has placed considerable burden on the technical ingenuity of South African water engineers. The solutions have succeeded in keeping abreast of water demand in the Greater Johannesburg Conurbation, but at an increasingly high cost to the environment. The existence of substantial water quality problems has arisen directly as a result of the human technical ingenuity needed to wrestle vast wealth from the bowels of the earth, and is likely to become one of the single most important challenges in the mid-term future. This will place major strain on the so-called second-order resources of society, the social adaptive capacity, or the ability of a given social entity to cope with rapid changes to water supply over time (Ohlsson, 1999; Ohlsson & Turton, 1999; Turton & Ohlsson, 1999; Allan, 2000, pp. 322–325). The lesson learned in this case is that societies need to constantly generate two distinct types of ingenuity if they are to continue to survive under conditions of endemic water scarcity. Technical ingenuity, or the capacity to solve problems through technological innovation, and social ingenuity, or the ability to design markets and create incentives to stimulate an adequate flow of technical ingenuity in the first place (Homer-Dixon, 2000, p. 22) are flip sides of the same coin. This is important because it is in this context that issues such as the cash-flow crisis confronting Rand Water, and the unpredictable nature of the outcome of the HIV/AIDS pandemic can best be understood. Both are undermining the capacity of society to generate new innovative solutions to problems of an increasingly complex nature. An understanding of the dynamics of this are largely absent from the contemporary water resource management literature.

The second major lesson learned is that the sustained supply of water to the Greater Johannesburg Conurbation over time has created a massive dependency on the continued supply of that water over time. The fact that the Rand Water supply area today sustains one of the largest concentrations of human beings that exist away from a river, lake or natural waterfront, suggests that future strategic outcomes are not guaranteed. In fact, there is some evidence to suggest that the increased need to develop ever more complex water transfer schemes, will result in an increased level of vulnerability to rapid changes in the system that sustains those transfers. These IBTs are embedded in society on the one hand, and the ecological systems needed to sustain those supplies on the other hand. Thus, uncertainties associated with over-abstraction and ecological decline, global climate change, deteriorating water quality caused by the closure of mines in the headwaters, and the unknown trajectory of the HIV/AIDS pandemic, can all be potentially magnified through the heavy reliance on technical and social ingenuity. This will place a major burden on engineers and political systems in the future as sustainable solutions are sought. An understanding of this aspect is also largely absent from the contemporary water resource management literature.

The third major lesson that can be distilled relates to the emerging issue of environmental-induced conflict. The key question relates to the causal linkage between water scarcity and conflict. Will endemic water scarcity drive conflict or cooperation in future? The fact that water issues already feature at a level of interstate interaction to the extent that an immature Hydropolitical Complex can be detected in Southern Africa, suggests that this question has not yet been fully answered. Initial indications are that water scarcity is acting as a driver of co-operation, but the jury is still out on the future outcome of this vexing issue. In this regard water supply to the Greater Johannesburg Conurbation is a useful litmus test for the rest of the world.

Conclusions

The water supply to a megacity like the Greater Johannesburg Conurbation is a study in complexity. The fact that Johannesburg exists at all has a lot to do with paleo-floodplain hydraulics, tectonic activities and the growth of primitive algal material that trapped the sediment and created the gold-bearing reef on which the city foundations rest. It is because of the existence of this gold, which has largely been mined out, that one of the largest human populations dependent on the ingenuity of water resource engineers exists in the Southern Hemisphere today. In the relatively short time-span of just over one century, the ecosystem around the Witwatersrand Ridge has been irreversible altered. In the wink of an eye in geological time, human ingenuity has transformed the Cradle of Humankind into a modern metropolis that is embedded in the largest man-made forest in the world. This raises serious questions about sustainability over time. More importantly it raises the issue of the link between second-order resources and social stability. Central to this is the emergence of the Southern African Hydropolitical Complex, with Johannesburg straddling the watershed between the two pivotal river basins in that Complex, the Orange and the Limpopo.

Acknowledgements

Rand Water is gratefully acknowledged for their support in providing many of the factual data used in developing this chapter. Professor Peter Ashton is acknowledged for his input with regard to acid rock drainage and water quality as applicable to South Africa. Carin van Ginkel is acknowledged for her various inputs with regard to water quality issues associated with highly eutrophic systems. The Stockholm International Water Institute is thanked for their generous financial support that allowed this chapter to be developsed in the first place. The authors alone accept responsibility for the contents of this paper.

Notes

1. See http://www.southafrica.net/heritage/heritage_places/cradle_1.cfm for more information.
2. A pivotal basin is one that is strategically important for the riparian states, which is reaching a point of closure where no more water is available for additional economic activities. See Turton (2003b, p. 79) for more information.
3. The Sterkfontein Heritage Site where the Cradle of Humankind is situated, is under threat from acid mine drainage entering the Tweefonteinspruit. This was first detected in 26 August 2002 and is now the focus of considerable environmental management concern. See Holtzhauzen (2004, p. 20) for more details.
4. For more background to the discovery of gold in Johannesburg, particularly with respect to the controversy about who was the first person to actually make the find, refer to Appelgryn (1984) and Cunningham (1987).
5. The first discovery of gold in what was to become Krugersdorp occurred in 1885 on the farm Kromdraai (translated as 'Crooked Bend' in a river). Krugersdorp has since been renamed as Mogale City.
6. When Johannesburg was officially founded, it was located on a piece of unaccounted for land (called Randjeslaagte) that had been left over when the original farms were demarcated (Appelgryn, 1984, p. 18). The reason for this was that there was no water on this land, making it useless for farming purposes. This saw a triangle of land some 240 hectares in extent being left over after farm delineation, to be taken over by the state. Significantly, this meant that when Johannesburg was first developed, a critical limiting factor became access to water. Associated with this was the pressing need for adequate sanitation. The first meeting of the Diggers Committee mandated Col. I.P. Ferreira, H.J. Morkel and Dr. H. Sauer to draw up a plan for sanitation (Appelgryn, 1984, p. 44). This saw the first comprehensive set of regulations being launched on 13 October 1887 (Appelgryn, 1984, p. 52).
7. Given that the currencies of the industrialized world at that time were based on the gold standard, with Britain as the leading industrial and economic power, this had strategic significance to the maintenance and expansion of the British Empire. For more background to this refer to Barber (1999, pp.11–12).
8. For more background to the Jamieson Raid see Seymour Fort (1908) and Longford (1982), both of which are considered to be authoritative.
9. The Anglo-Boer War was a resource war triggered by the British desire to gain control over the gold that had been discovered in the Zuid Afrikaanse Republiek (ZAR) (Barber, 1999, pp. 11–12). For more background see Farwell (1999), Nasson (1999) and Lee (2002).
10. Barber (1999, pp. 10–11) notes that despite the fact that there were two separate and independent Boer Republics in southern Africa at the time, the dominant belief in London was that this was firmly a British sphere of influence. So for example, when diamonds were discovered in Griqualand in the late 1860s, claims by the Boer Republics were quickly countered and the British absorbed the diamond-bearing territory into the Cape Colony. This laid the foundation to the belief that the subsequent Anglo-Boer War was a resource war.
11. The establishment of concentration camps in South Africa during the Anglo-Boer War was a direct result of Lord Roberts scorched earth policy (Van Rensburg, 1980, p. 37). This saw the removal of Boer women and children from the farms in the belief that this would prevent their support to the men who were out on Kommando fighting a guerrilla war (Evans, 1982). This thinking was based on earlier experiments with concentration camps in Cuba (Spies, 1977, p. 148), known as the Weyler Method (Spies, 1977, p.148; Van Rensburg, 1980, p. 38). For further accounts of the concentration camps see Farwell (1999), Nasson (1999), Raath (1999) and Lee (2002). For the linkage between concentration camps and imperialism, see Porch (2000).
12. For more background to the social stratification caused by the discovery of gold in Johannesburg, refer to Barber (1999, p. 12).
13. The Afrikaner is a much-misunderstood nation. The name Afrikaner literally means 'of Africa' or 'African' (Turton, 1999). They speak a language that was originally derived from Dutch, but which has been Africanized and is spoken on no continent other than Africa (with a small exception in Argentine where some Boer trekkers settled and established a small but non-viable Afrikaner community of herdsmen). The Afrikaner is an African in every sense of the word and regard themselves as the so-called White Tribe of Africa (Harrison, 1982), having engaged in an anti-imperial struggle during the Anglo-Boer War and therefore largely sympathetic to the anti-colonial aspirations of other African nations. Their 'Africaness'

is being given expression in post-Apartheid South Africa where they are taking their rightful place in what has become known as the Rainbow Nation.

14. The deepest mine is the Western Deep Levels, which is on the Witwatersrand complex in South Africa, with a depth in excess of 3500 metres. Planned operations are projected to take mining to a level of some 4117 metres, with research being conducted at the Council for Scientific and Industrial Research (CSIR) in South Africa attempting to go even beyond that depth (*Financial Times*, 1999).

15. The area supplied today by Rand Water has a population in excess of 11 million people, or a quarter of the total population of South Africa, who produce 34% of the country's GDP, or 10% of the GDP of the entire African continent (Gauteng Economic Development Agency, 2004).

16. The Zuurbekom dolomites yield high quality groundwater that does not need treatment before being entered into the overall supply system of present day Johannesburg. Yielding an annual average of 30.5Ml/d when first developed in 1899, this tapered off to 25 Ml/d in 1978 and 12Ml/d in 2003. This water is still of an exceptionally high quality.

17. By 1889 there were 711 crushers in operation. A year later this had grown to 6000 crushers. By 1898 some 7.3 million tons of ore was being milled, translating into a demand for water in the order of 14.6 tons of water.

18. The Braamfontein Spruit is one of the rivers that is now encased in the stormwater sewers of Johannesburg. A tributary of the Limpopo river basin, the Braamfontein Spruit discharges into the Hartebeespoort Dam, which is one of the most polluted dams on the continent of Africa today.

19. This is translated literally as 'Four Fountains'.

20. This is translated literally as 'Elephant Marsh', referring to the original wetlands of the area.

21. The Pipelines of Power thesis suggests that political power is derived from the development and control over water pipelines in semi-arid areas. This is based on the assumption that because water is scarce, it gives rise to economic privilege. Control over the supply of water thus translates to political power, leading to a degree of resource capture and ecological marginalization to use the concepts central to the work of Homer-Dixon (1991, 1994a, 1994b, 1995, 1996, 2000).

22. For a concise history of Rand Water see Tempelhoff (2003).

23. The Rand Water Board was modeled on the constitution of the Metropolitan Water Board of London, now known as Thames water, which was also formed in 1902.

24. These talks were driven by the desire to establish a large steel factory in Pretoria. Initial talks were stalled but again revived in 1934 when the Vaal Dam Project got underway.

25. There is a strong parallel here with the construction of the Hoover Dam during the Great Depression in the USA. Part of the New Deal, this thinking saw hydraulic infrastructure development as a sound hedge against the debilitating effects of economic recession. This thinking permeated the Hydraulic Mission of the western industrialized nations for the major portion of the 20th century. For more background see Reisner (1993) and Allan (2000).

26. For a more detailed assessment of this refer to Turton *et al.* (2004a).

27. This refers to the heavily industrialized triangle with Pretoria, Witwatersrand and Vereeniging on the respective corners.

28. This scheme took water from the Thukela river basin on the eastern escarpment, pumping it over the Drakensburg Mountains using surplus electricity from the national grid, and then reclaiming a significant portion of that energy when needed by gravity feeding the water into the upper reaches of the Vaal river. The first water was pumped into the Sterkfontein Dam in October 1974.

29. Phase 2 of the LHWP consists of the construction of the Mashai Dam shown in Fig 3 and associated water delivery infrastructure. Phase 3 consists of the construction of Tsoelike Dam and associated delivery infrastructure.

30. For more details of the feasibility study of the next phase of the Thukela Water Project see http://www.dwaf.gov.za/thukela/feasibility.htm (accessed 13 July 2004).

31. For a deeper understanding of the linkage between HIV/AIDS and water, refer to Ashton & Ramasar (2002). In addition to this, patients on anti-retroviral treatment need clean water if they are not to have their health compromised further. This is a problem that seems to receive surprisingly little attention in the contemporary water resource management literature.

32. Professor Peter Ashton is a water quality specialist at the Council for Scientific and Industrial Research (CSIR) in Pretoria, South Africa.

33. See Holtzhauzen (2004, pp. 18–20) for some indication of the nature of the problem.

34. Carin van Ginkel is a specialist scientist at Resource Quality Services in the South African Department of Water Affairs and Forestry (DWAF).

35. For example, see the study commissioned by the Department of Water Affairs and Forestry in 2003 that is being done by Southern Waters (DWAF, 2003a, 2003b). The final report is not yet available at the time of writing.

References

Allan, J. A. (2000) *The Middle East Water Question: Hydropolitics and the Global Economy* (London: IB Tauris).

Appelgryn, M. S. (1984) *Johannesburg 1886–1889: Origins and Early Management* (Pretoria: University of South Africa).

Ashton, P. & Ramasar, V. (2002) Water and HIV/AIDS: some strategic considerations in Southern Africa, in: A. R. Turton & R. Henwood (Eds) *Hydropolitics in the Developing World: A Southern African Perspective*, pp. 217–235 (Pretoria: African Water Issues Research Unit (AWIRU)).

Barber, J. (1999) *South Africa in the Twentieth Century* (Oxford: Blackwell Publishers).

Chapman, K. M., Tucker, R. F. & Kidger, R. J. (1986) The Klerksdorp goldfield, in: E. S. A. Antrobus (Ed.) *Witwatersrand Gold—100 Years*, pp. 173–197 (Pretoria: The Geological Society of South Africa).

Cunningham, A. (1987) *The Strubens and Gold* (Johannesburg: Ad. Donker Ltd).

Davie, L. (2003) Joburg's First Trees. Available at http://www.joburg.org.za/2003/aug/aug25_trees.stm

DWAF (2003a) Hartebeespoort Dam Remediation Project (Phase 1), *Inception Report* (Pretoria: Department of Water Affairs and Forestry).

DWAF (2003b) Hartebeespoort Dam Remediation Project (Phase 1), *Deliverables Report: Task 1 (Literature Review)* (Pretoria: Department of Water Affairs and Forestry).

Evans, M. M. (1999) *The Boer War: South Africa 1899–1902. Osprey Military* (Mechanicsburg: Stackpole Books).

Farwell, B. (1999) *The Great Boer War* (London: Wordsworth Editions).

Financial Times (1999) Down, down, deeper and down, *Financial Times Weekend*, June 5. Available at http://deepmine.csir.co.za/Press%20Clips/clip02.htm (accessed 5 July 2004).

Gauteng Economic Development Agency (2004) Gauteng Overview. Available at http://www.geda.co.za (accessed 5 July 2004).

Harrison, D. (1982) *The White Tribe of Africa: South Africa in Perspective* (Berkeley: University of California Press).

Holtzhauzen, L. (2004) Decanting minewater: solving a 100-year-old problem, *Water, Sewage and Effluent*, 24(4), pp. 18–20.

Homer-Dixon, T. F. (1991) On the threshold: environmental changes as causes of acute conflict, *International Security*, 16(2), Fall, pp. 76–116.

Homer-Dixon, T. F. (1994a) Environmental changes as causes of acute conflict, in: R. K. Betts (Ed.) *Conflict after the Cold War: Arguments on Causes of War and Peace* (New York: Macmillan).

Homer-Dixon, T. F. (1994b) Environmental scarcities and violent conflict: evidence from cases, *International Security*, 19(1), pp. 5–40.

Homer-Dixon, T. F. (1995) The ingenuity gap: can poor countries adapt to resource scarcity?, *Population and Development*, 21(3), pp. 587–612.

Homer-Dixon, T. F. (1996) Environmental scarcity, mass violence and the limits to ingenuity, *Current History*, 95, pp. 359–365.

Homer-Dixon, T. F. (2000) *The Ingenuity Gap* (London: Jonathan Cape).

Lee, E. (2002) *To the Bitter End: A Photographic History of the Boer War 1899–1902* (Pretoria: Protea Book House).

Longford, E. (1982) *Jamieson's Raid: the Prelude to the Boer War* (London: Weidenfeld & Nicholson).

Mitchell, P. (2002) *The Archaeology of Southern Africa* (Cambridge: Cambridge University Press).

Nasson, B. (1999) *The South African War 1899–1902* (London: Arnold).

Ohlsson, L. (1999) *Environment, Scarcity and Conflict: A Study of Malthusian Concerns* (Göteborg: Department of Peace and Development Research, University of Göteborg).

Ohlsson, L. & Turton, A. R. (1999) The turning of a screw. Paper presented at the Plenary Session of the 9th Stockholm Water Symposium Urban Stability through. Integrated Water-Related Management, Stockholm International Water Institute (SIWI), Sweden, 9–12 August. Available as *MEWREW Occasional Paper No.19* at http://www.soas.ac.uk/Geography/WaterIssues/OccasionalPapers/home.html and from http://www.up.ac.za/academic/libarts/polsci/awiru

O'Keeffe, J., Uys, M. & Bruton, M. N. (1992) Freshwater systems, in: R. F. Fuggle & M. A. Rabie (Eds) *Environmental Management in South Africa* (Johannesburg: Juta & Co).

Porch, D. (2000) *Wars of Empire* (London: Cassell).

Raath, A. W. G. (1999) *The British Concentration Camps of the Anglo Boer War 1899–1902: Reports on the Camps* (Bloemfontein: The War Museum). Available from Thorrold's Africana Books.

Rabie, M. A. & Day, J. A. (1992) Rivers, in: R. F. Fuggle & M. A. Rabie (Eds) *Environmental Management in South Africa* (Johannesburg: Juta & Co).

Reisner, M. (1993) *Cadillac Desert: The American West and its Disappearing Water,* revised edn (New York: Penguin).

Seymour Fort, G. (1908) *Dr. Jamieson* (London: Hurst & Blackett Ltd).

Spies, S. B. (1977) *Methods of Barbarism: Roberts, Kitchener and Civilians in the Boer Republics January 1900–May 1902* (Cape Town: Human & Rousseau).

Tempelhoff, J. W. N. (2003) *The Substance of Ubiquity: Rand Water 1903–2003* (Vanderbijlpark: Kleio Publishers).

Turton, A. R. (1999) Statutory Instruments for the Maintenance of Ethnic Minority Interests in a Multi-Cultural Community: The Case of the Afrikaners in South Africa. (Translated into Russian as: Pravovye mery zashchity interesov etnicheskih menshinstv v mnogonatsionalnom obshchestve: afrikanery Yuzhnoy Afriki (Legal measures of defending interests of ethnic minorities in a multinational society: The Afrikaners of South Africa), in: N. I. Novikova & V. Tishkov (Eds) *Folk Law and Legal Pluralism.* Proceedings of the 11th International Congress on Folk Law and Legal Pluralism, August 1997, Moscow pp. 38–47 (Moscow: Institute of Ethnology and Anthropology).

Turton, A. R. (2000) Precipitation, people, pipelines and power: towards a political ecology discourse of water in Southern Africa, in: P. Stott & S. Sullivan (Eds) *Political Ecology: Science, Myth and Power* (London: Edward Arnold).

Turton, A. R. (2003a) An Introduction to the hydropolitical dynamics of the Orange river basin, in: M. Nakayama (Ed.) *International Waters in Southern Africa* (Tokyo: United Nations University Press).

Turton, A. R. (2003b) Environmental security: a Southern African perspective on transboundary water resource management, *Environmental Change and Security Project Report.* The Woodrow Wilson Centre, issue 9, Summer, pp. 75–87 (Washington, DC: Woodrow Wilson International Center for Scholars).

Turton, A. R. (2004) The evolution of water management institutions in select Southern African international river basins, in: C. Tortajada, O. Unver & A. K. Biswas (Eds) *Water as a Focus for Regional Development* (London: Oxford University Press).

Turton, A. R. & Earle, A. (2005) Post-apartheid institutional development in selected Southern African international river basins, in: C. Gopalakrishnan, C. Tortajada & A. K. Biswas (Eds) *Water Institutions: Policies, Performance & Prospects*, pp. 154–173 (Berlin: Springer).

Turton, A. R. & Ohlsson, L. (1999) Water scarcity and social adaptive capacity: towards an understanding of the social dynamics of managing water scarcity in developing countries. Paper presented at the Workshop No. 4: Water and Social Stability of the 9th Stockholm Water Symposium Urban Stability through Integrated Water-Related Management, Stockholm International Water Institute (SIWI), Sweden, 9–12 August. Available as *MEWREW Occasional Paper No. 18* at http://www.soas.ac.uk/Geography/WaterIssues/OccasionalPapers/home.html and http://www.up.ac.za/academic/libarts/polsci/awiru

Turton, A. R., Meissner, R., Mampane, P. M. & Seremo, O. (2004a) *A Hydropolitical History of South Africa's International River Basins.* Report to the Water Research Commission (Pretoria: Water Research Commission).

Turton, A. R., Ashton, P. J. & Cloete, T. E. (2004b) An introduction to the hydropolitical drivers in the Okavango river basin, in: A. R. Turton, P. J. Ashton & T. E. Cloete (Eds) *Transboundary Rivers, Sovereignty and Development: Hydropolitical Drivers in the Okavango River Basin*, pp. 9–30 (Pretoria and Geneva: African Water Issues Research Unit (AWIRU) and Green Cross International).

Van Rensburg, T. (1980) *Camp Diary of Henrietta E.C. Armstrong: Experiences of a Boer Nurse in the Irene Concentration Camp 6 April–11 October 1901* (Pretoria: Human Sciences Research Council) (HSRC)).

Venter, P. (2004) New hope for troubled waters: the Hartebeespoort dam test case, *The Water Wheel,* 3(1), January/February, pp. 16–19.

WCD (2000) *Dams and Development: A New Framework for Decision-Making* (London: Earthscan).

Wentzel, R. G. (2001) *Limnology, Lake and River Ecosystems* (New York: Academic Press).

Werdmüller, V. W. (1986) The Central Rand, in: E. S. A. Antrobus (Ed.) *Witwatersrand Gold—100 Years,* pp. 7–47 (Pretoria: The Geological Society of South Africa).

Water Management in Metropolitan São Paulo

B. P. F. BRAGA, M. F. A. PORTO & R. T. SILVA

Introduction

Considerations about the long-term prospects of water resources development and the balance between alternatives to expand capacity and management of demand are attributes that have been associated with a more recent view of water resources management, and this has been pursued by water managers worldwide. Initiatives to control pollution at the source, as opposed to the conventional position of extending the structural capacities to treat the effects, are elements increasingly applied to the case of urbanized basins. However, this development of the management culture is not performed uniformly covering all sectors and uses. In the case of the Metropolitan Region of São Paulo, the São Paulo State laws enacted to protect water sources represent an important contribution to this new management approach.

In densely urbanized basins a new concept, Total Urban Water Management (TUWM) is proposed. In this new approach, integration is applied indistinctly to the sectorial vectors (combining different water uses) and to the territorial vectors, in the sense of horizontally cutting across different jurisdictions on the territory. With regard to sectorial integration, the TUWM of urbanized basins includes the need to communicate with sectors that do not use the resources, such as municipal management, housing and urban transport, as well as the multiple uses of water resources themselves (industrial, public water supply, sewerage, storm drainage). As a result of the recognition of these dimensions of institutional integration/communication there is a structural connection between water resources management and the regional/metropolitan planning, the latter with jurisdiction on the common public functions that extrapolate the sphere of the water resources itself.

In the following institutional analysis, concerning the national and state water resources systems, rather than a formal description, the study seeks to identify the elements of sectorial and territorial integration compatible with the institutional concepts of these systems. From the viewpoint of territorial communication, the main element sought is the institutional/legal propensity to intergovernmental cooperation between different territorial aggregations, considered the three basic units of the Brazilian Federation, the Union, the states and the municipalities. From the viewpoint of sectorial communication, the potentials are analysed based on the figures of inter-institutional cooperation in a same sphere of territorial aggregation, seeking solutions that will make public functions more effective as a whole. These two viewpoints converge, and in many cases are practically indistinguishable, since the jurisdictions or some public functions are intrinsically associated with a sphere of public administration in particular.

A major challenge faced by water managers of the large cities in the developing world today is the provision of water infrastructure to an extremely fast growing system. Urban population has increased, in some cities (e.g. São Paulo, Mexico City, Bombay), almost tenfold in the last 50 years. The needs for water supply, sewage collection and appropriate disposal, urban flood control and solid waste disposal have similarly increased. On the other hand, financial resources have not been put in place at the required rate. The consequence of this situation is the well-known contamination of urban waters with the associated public health hazards and frequent flooding of the riparian (poor) population.

More recently, with an improvement in the level educational of the urban population in the developing world, this situation is beginning to change. Citizens are exercising their rights by organizing themselves into associations, which in turn press the politicians towards providing more funds to restore the urban environment. In some countries where the water legislation allows for water organizations and water users to participate in the decision process (e.g. Brazil and Mexico) positive results are becoming apparent.

This study describes the situation of the Metropolitan Region of São Paulo where a century of mismanagement has required, in the last decade, massive investments to improve the quality of the urban environment. It represents a contribution to the discussion of the water management in megacities as a case study of hope. Hope in good management of governments in less developed countries and hope in the participation of citizens in the water management decision process. The concept of TUWM is proposed and the means for its implementation described.

Background

The São Paulo Metropolitan Area (MRSP) includes the city of São Paulo and a further 3 municipalities, and together they occupy an area of approximately 8000 km², 1500 km² o

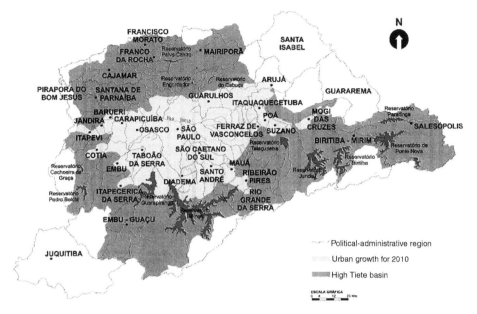

Figure 1. The Metropolitan Region of Sao Paulo (MRSP) and the upper Tiete river basin

which are highly urbanized. The current population of this region is 18 million inhabitants, with a forecast of 20 million inhabitants by the year 2010. This region (Figure 1) is the largest urban concentration and industrial complex of Latin America. This is the most important area for the production of industrial goods in Brazil, since its Gross Domestic Product (GDP) in 1997 reached US$147 billion, corresponding to approximately 18% of the Brazilian total. The industrial sector is very significant, both in creating jobs and income. The service sector is also growing and in recent years has developed and created the most new jobs in the region.

The MRSP is located in the upper Tiete river basin, which at Edgard de Souza dam has a drainage area of approximately $4000 \, km^2$. Gentle slopes (of the order of $0.17 \, m/km$) characterize a meandering Tiete river. From its headwaters until Edgard de Souza dam the Tiete river flows through 161 km, having as its main tributaries the Tamanduatei and the Pinheiros rivers. São Paulo Metropolitan region is geographically located just in the Capricorn Tropic line but its high altitude (750 m) changes the climate of the region to Cwb (Koppen classification) that is, temperate climate with dry winter. Annual average temperature is 19°C, varying from a minimum of 15.5°C to a maximum of 25°C. Due to the proximity of the region to ocean its average humidity is high (80%). Its complex relief and the sea breeze bringing moist air in the afternoon facilitate the occurrence of convective cells in the summer, producing floods in small basins. Due to the geomorphologic characteristics of the Tiete river, floods have always been a problem in the region.

This high population and industrial growth rate in the last 30 years created an imbalance between water supply and demand. Furthermore, the growing imperviousness of urban soil magnified a basin flooding problem. The upper Tiete river basin is being urbanized at a very high rate upstream of Penha dam and is almost completely urbanized downstream of this dam. Flood hydrographs for the upstream basin show a slow rising limb with moderate peaks, typical of rural areas, while downstream of that dam, flood hydrographs are

Table 1. Growth of the urban area

Year	Population		Urbanized area	
	Million of inhabitants	Annual growth rate (%)	Related to the basin (%)	Increment (%)
1905	0.3	10.3	0.6	433
1930	1.1	8.7	3.2	172
1954	3.5	8.3	8.7	154
1973	9.3	4.1	22.1	63
1985	14.8	1.1	36.0	21
2004	18.0		43.8	

typically urban. Table 1 shows the urban growth in the region in the last 100 years. At the beginning of the century São Paulo was a small city, with 300 000 inhabitants, who occupied 0.6% of the total area of the basin, and in less than a century it has transformed itself into a megalopolis of 18 million inhabitants.

Industrialization and urban growth occurred at the cost of a higher generation of sewage and the need to look for water supply sources in neighboring watersheds away from the demand centre. Water supply demands have grown exponentially and today the metropolitan area water supply utility, SABESP, faces tremendous challenges in both technical and financial terms to supply these growing demands.

Basin Occupation

The upper Tiete river basin has an urbanized area covering approximately 37% of the basin area, and despite the current process of reduction in the population growth rates, this is not reflected in contained expansion of the urban sprawl. The expulsion of the low-income population to the outskirts of the cities worsens environmental degradation because of disordered expansion and lack of adequate urban infrastructure, generating the consequent problems of occupation of water sources and floodplain protection areas, in addition to the need to expand the water supply, sanitary sewerage and solid waste collection systems. This has a very serious consequence for the region, namely the need to continue investing in expanding the urban infrastructure at rates higher than those of the overall population growth. The downtown area becomes less populated and its already installed and consolidated infrastructure becomes increasingly idle, with swelling outskirts that painfully await a time when the system can increase its investments and bring them the necessary basic infrastructure.

The highest population growth rates are in the water source protection areas. Uncontrolled urban occupation in the protected areas is the greatest threat to the sources. This occupation produces domestic sewage, solid waste and a non-point urban pollution load. Bulk water quality is seriously affected, resulting in higher treatment costs and a threat of reducing the quality of water to be distributed to the population due to the possible presence of toxic substances associated with urban pollution.

It should be stressed that the loss of any of the surface water sources currently used to supply the Metropolitan Region of São Paulo will create irretrievable problems for the region supply system. The main problem relating to the protection of sources is the fact that the protection of these areas, with regard to disciplining the use and occupation of the

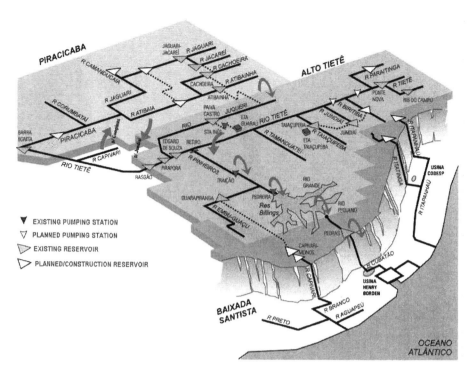

Figure 2. The Upper Tiete river basin and surroundings. *Source:* Braga (2000).

land, is not the task of the system that manages the water resources, but rather of the municipalities belonging to the respective producing basin. Only an integrated management system can shed some light on the solution to this problem. This is one of the main points where it is urgently necessary to attach water resources management to urban management of the territory.

Inter-basin Water Transfers

One of the first cases of inter-basin water transfer in the South American continent took place in the upper Tiete river basin. The first Brazilian hydropower plant was inaugurated in 1901 in the Tiete river, downstream from the city of São Paulo at the site of Parnaiba dam (currently renamed Edgard de Souza dam in Figure 2). Since it was impossible to build a large dam and reservoir without impacting the city of São Paulo, in 1908 a second regulating dam was built, Guarapiranga on the Embu-Guaçu river, a tributary of Pinheiros river (Figure 2). In 1912 the power generated by Parnaíba plant was raised to 16 MW. Due to a drought in 1924, there were energy supply problems in the region, which allowed the power utility Light & Power Co. to start a very creative project, the Serra do Mar Project, in 1927. The purpose of this project was to invert the natural flux of the waters of Tiete river to the maritime slopes of Serra do Mar, and using a new hydropower plant, Henry Borden Plant, at the foot of the mountain range (750 m drop) generate plenty of power for the city of São Paulo and surroundings. This system involved raising the height of Edgard de Souza dam, and by successive pumping in Traição and Pedreira stations store water in Billings reservoir and thence to the Pedras reservoir and finally transfer the waters of the Tiete river to the Cubatão river in the Baixada Santista region (Figure 2). Thus, the installed capacity of the region was increased from 16 MW to 480 MW in early 1950s, which enabled the speedy industrialization and urbanization of the region. The lack of wastewater collection and treatment in the upper Tiete basin, in the following years caused the complete eutrophication of Billings reservoir and other important water bodies.

The disordered growth of the MRSP, notably beginning in the 1950s, made the water supply system in the region inadequate, with successive situations of lack of water and rotations. In 1966, the construction of a system to transfer the waters of the rivers forming the Piracicaba river to the upper Tiete basin began. The works as a whole use water resources from the headwaters of the Piracicaba rivers by means of dams located in rivers Cachoeira, Atibainha and Jaguari/Jacare, as well as damming the Juqueri river, which already lies in the upper Tiete basin. The Cantareira System supplies approximately 33 m³/s, corresponding approximately to 60% of all of the water consumed in the São Paulo Metropolitan Area.

The reservoir system (Jaguari, Cachoeira, Atibainha, Paiva Castro and Águas Claras) are located at different levels and interconnected in such a manner that, from the Jaguari to the Paiva Castro, waters go through Cachoeira and Atibainha by gravity, and reach Santa Inês pumping station, where the full volume produced is pumped to the Aguas Claras Reservoir, already at the upper Tiete basin at the top of the Serra da Cantareira. From this reservoir the waters are treated at the adjacent Treatment Plant of Guaraú, and then sent to the MRSP for consumption.

In the 1960s it was unusual for the population to participate in decision-making processes concerning water resources. Even so, in the case of the Cantareira system, considering its potential impacts on the source Piracicaba river basin, meetings took place between the Secretary of Civil Works of the State of São Paulo and the basin community in the city of Piracicaba. The meetings at Piracicaba were the embryo of the present Inter-municipal Consortium of the Piracicaba basin. This consortium of municipalities is politically highly influential in the decision-making process concerning water resources in the basin, and shows how the community can begin from basics to demand their own regional solutions.

The current situation of the Piracicaba river basin is completely different from that when the system was implemented at the beginning of the 1970s. The basin became urbanized and currently 8.5% of the state's population lives in its 12 400 km², mainly in urban areas (87%). Complicating this situation even further, irrigation in this region has considerably increased and the maintenance of minimum discharges in adverse climatic situations (low flows) is becoming increasingly difficult. A new water management model is in place with a very active Piracicaba river basin committee. Under this new arrangement SABESP had to renew its authorization to transfer water to the MRSP. This authorization was given by the National Water Agency and the Department of Water and Power of the State of São Paulo under the express requirement that water should be provided to the Piracicaba river basin to meet their basic demands.

Flood Control

The solution of the problem of urban floods can be summarized as allocation of space. During the rainy season, rivers carry more water and therefore need space to transport it.

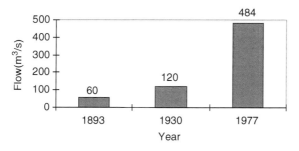

Figure 3. Design flows of the Tamanduateí river at Glicério Station

The space thus occupied is called the river floodplain. The only way to control the floods is, thus, to provide space that will be used by floodwaters. This can be achieved by preserving the floodplain areas, or by creating new detention dams and reservoirs.

Space, lots and real estate are valued assets in the urban areas. Removing the floodplain occupants, as a corrective measure, is much more expensive, from several standpoints, than preventing it from being occupied. Equally, 'creating' spaces to store the excess, as in the case of the flood detention structures, is also expensive, especially as urbanization becomes denser.

One of the recurring problems of floodplain occupation in the Metropolitan Region of São Paulo in general, and in the Municipality of São Paulo in particular, is the model of implementation of valley bottom avenues. If on the one hand they have the advantage of increasing the benefits of public investment in drainage and roads, on the other hand they induce a conventional, dense pattern of land use and occupation, which at the same time contributes to increasing floods and is more vulnerable to their consequences. It should be noted that this is also a problem whose control depends on land-use planning and occupation. These issues are managed by the municipalities and as in the occupation of source protection areas, this integration of spheres of power is essential to control the process.

It is essential for the upper Tiete basin that the floodplain occupation upstream from the Penha Dam be contained, and also that all restriction discharges advocated by the Upper Tiete Basin Macrodrainage Plan be maintained They are called restriction discharges because they delimit the maximum discharge capacity through the existing channels. These discharges will only be feasible when strong policies are implemented to control the process of occupying and rendering floodplains impervious. In a more comprehensive context for the upper Tiete basin, this means the need for strong interaction between the housing policies and the water policies.

Due to inadequate land use in the basin, floods are becoming more severe in small creeks and the problem is being transferred to the Pinheiros and Tiete river basins. The situation has become extremely critical since the main freeways of São Paulo are located exactly at the banks of the Tiete and of the Pinheiros rivers. During the wet season (December to March) the population is frightened, and radio, television and newspapers give much coverage of the problem. It is clear that any hydraulic work done in the Tiete or Pinheiros will be inefficient in the future if flood control is not carried out in the contributing basins. Figure 3 shows the design flows of the Tamanduateí river at Glicério gauging station. It clearly shows the importance of changing from end-of-pipe to source control solutions. In less than a century design flows have increased more than fivefold.

Table 2. Detention volumes in the contributing basins of the Tiete River ($T = 25$-year)

Basin	DA (km^2)	Total volume (m^3)	Detention basins	
			Planned	Built/operation
Tamanduateí	330	7.7×10^6	43	16
Pirajussara	72	2.1×10^6	16	05
Aricanduva	100	2.2×10^6	11	08
Médio Juqueri	263	3.1×10^6	26	–
Canal de Circunvalação	33	1.1×10^6	03	–
Baquirivu Guaçu	136	3.5×10^6	31	–

Source: Canholi, (2003).

DAEE, the agency responsible for water management in the State of São Paulo, is currently implementing a very modern urban drainage master plan. This master plan is considering modern technology by using detention basins and canals in a coordinated fashion. The urban drainage master plan has been developed using three main principles: urban drainage is a regional issue; preference should be given to store water in the drainage basin; urban drainage is multidisciplinary. Heavy civil works are in progress in the Tiete river to increase its capacity to nearly 2000 m^3/s. At the same time restrictions have been set at the main tributaries outlets in order to guarantee that the design flows will not be surpassed in the future.

Table 2 shows the detention volumes to be stored in the main tributary basins of the Tiete river in order to achieve protection of the 25-year return period flow. These detention basins operate on-line and off-line and have proven to be effective during the flood season in the Aricanduva basin. A significant problem faced by megacities in developing countries in using these detention basins is the amount of debris and solid wastes that have to be removed after a flood event. This fact emphasizes the importance of a multidisciplinary approach in which solid waste management sector must be involved. In addition, there are different jurisdictional competences with the municipality being responsible for solid waste management and the state for water management.

Wastewater Collection and Treatment

Wastewater collection and treatment is a reoccurring problem in the MRSP which began in the early 1950s. The first wastewater master plan at that time included five wastewater treatment plants located on the Tamanduateí and Tiete river banks. Although the master plan was very well executed, its implementation was never considered by the decision-makers of the municipalities involved.

During the 1980s, the sanitary situation of the Tiete river basin was intolerable. Environmental movements and the press (notably the AM radio stations) began a popular movement for the environmental recovery of the river. Groups in the Billings reservoir area were dissatisfied with the transfer of São Paulo wastewaters upstream, and the deterioration of water quality in the reservoir. Popular pressure on legislators culminated with a transitory provision in the 1988 constitution of the State of São Paulo to prohibit the reversion of the Tiete waters until 1992. Before 1992 the operational practice involved partitioning discharge at the confluence with the Pinheiros river, 50% being pumped upstream to subsequent hydropower generation at Henry Borden and 50% naturally

Table 3. Sewage treatment plants of the Tiete Project in 2004

STP sewage treatment plant	Installed capacity (L/s)	Population
Barueri	9500	4 460 000
Suzano	1500	720 000
ABC	3000	1 400 000
Parque Novo Mundo	2500	1 200 000
São Miguel Paulista	1500	720 000
Total	18 000	8 500 000

Source: SABESP (2004).

flowing downstream inland. Despite the large volume of Billings reservoir (1.2 billion m^3), its central part became a large anaerobic oxidation pond.

After the transitory provision of the 1988 state constitution, all of the Tiete waters have flowed down the river's natural course. As a consequence the Henry Borden hydropower plant has been shut down. This represents an annual cost of approximately R$200 million (US$70 million). In addition, the problem of anoxic conditions in Billings reservoir have been transferred to the reservoirs in the middle branch of the Tiete river. The parties who suffer the impacts, led by environmental movements in the middle Tiete region, are currently pressuring and demanding means to compensate for environmental and economic losses resulting from the existing operational scheme.

A major project started in the 1980s (Tiete Project) aimed to clean up the waters of the Tiete river basin. This was the result of a massive media campaign towards reversing the daunting water quality situation of the Tiete waters. The generated effluent in the basin amounts to 900 ton BOD/day distributed: 36% non-point sources, 12% industrial, 52% domestic. The project involves the construction of collectors, mains, pumping stations and five sewage treatment plants at secondary level. The final configuration of the project by the year 2015 will provide sewage treatment for a population of 18 million and treat an effluent of 53 m^3/s. The remaining effluent load at the end of the project is estimated at 250 ton BOD/day. The situation of the STP in 2004 is shown in Table 3.

Water Supply

The Metropolitan Region of São Paulo demands 67 m^3/s of drinking water to supply its 18 million inhabitants. The water supply utility operates several production systems and three of them are large systems, producing 90% of the total production. The largest system is the Cantareira System, described before, with a production capacity of 33 m^3/s. The second largest system is the Guarapiranga and the Billings reservoir. This system produces 14 m^3/s, being responsible for supplying almost 20% of the entire demand. Their watersheds are located on the edge of the urban area of the city of São Paulo. Both gradually suffered an unplanned and disorganized occupation process during the last 30 years due to the implementation of an industrial district near the reservoir. There is a population of 1.5 million inhabitants living within the watersheds, the great majority of whom are low-income families. Urban areas occupy more than 15% of the total watershed area.

The Guarapiranga watershed has been a 'protected' area since the 1970s, as a water supply source area. There were laws protecting the area, with special provisions to prohibit any form of occupation. Unfortunately, as with most laws that are 'too' restrictive, it did not work. Poor people with no opportunity of being able to afford to buy a home, began invading this large empty space, conveniently close to the city and to the industrial district where there were plenty of jobs. Enforcement of the law was impossible. In just a few years thousands of people moved to the region and no services were provided since they were in an illegal situation as invaders of a protected and public area. Living conditions were very difficult and huge pollution loads from sewage and solid waste began being dumped into the reservoir. In 1990, algae blooms became the first nuisance observed by the water supply company. An intensive water quality degradation process was under development due to discharges of untreated domestic waste. An immediate remediation project was needed. In 1992, São Paulo water managers decided that the Guarapiranga system was a precious water source for the city of São Paulo due to its yield and the close location. A US$336 million seven-year project to recuperate and maintain the water quality of the reservoir then began. A World Bank loan of US$119 million together with a collateral investment of the State of São Paulo of US$217 million allowed the implementation of the project, which was finished during 1999. The success of this project led the World Bank to finance another project to protect the other supply systems.

The Alto Tiete System supplies $10\,m^3/s$ to the MRSP. The system is the only one that will support further expansion up to $15\,m^3/s$. Five reservoirs, linked by channels, tunnels and pumping stations, bring water to the treatment plant near São Paulo. Its watershed is under intense pressure due to urban expansion. The degradation of water quality may compromise the system in the near future.

The water supply system of MRSP is operated by a single public utility, SABESP. This is a very complex operational system. It involves seven subsystems and eight water treatment plants, 1472 km of mains, 159 pumping stations, 198 urban reservoirs and 25 000 km of distribution lines. Water sources are located far from the areas of greatest demand and are constantly under the threat of contamination. This is due to illegal settlements in the contributing drainage basins. The cost of drinking water treatment has greatly increased in the last five years. As expected, the highest increase, 133%, occurred in the Guarapiranga-Billings system. In the Alto Tiete system the cost increased 20% and in the Cantareira system it increased 27%. To produce $1000\,m^3$ of drinking water from the Guarapiranga system the cost is currently US$18, whereas from the Alto Tiete system it is US$11 and from the Cantareira system is US$3. This is the cost of pollution and lack of protection of the supply systems.

It will be very difficult to expand the water supply system in the MRSP. Water transfer from neighbouring basins will be necessary but it will impose large political and social costs. Together with the expansions, the implementation of non-structural measures such as conservation and reuse will be essential.

Total Urban Water Management

From an institutional point of view the situation in the MRSP is very complex because the municipality is in charge of the land-use planning, urban housing and transportation, while the state is responsible for water resources management. Fortunately, during the last decade the government was very receptive to advice from the professionals in the sector. As a result, the municipality today passes construction codes that require coping with floods at the lot level.[1] At the same time the state is investing massively in the Tiete project to improve the river's water quality.

It seems that efforts should be made in the direction of moving away from the traditional model described above to the one where urban water management should be undertaken at the intersection of the water supply and environmental sanitation sector, urban housing and transportation sector and urban drainage sector. It would be naive to think that, from an institutional point of view, a single institution would take care of all the tasks related to the three sectors described. Certainly, a system should be established in such a way that the different sectors would act in a coordinated fashion. The river basin as a planning unit is a necessary but certainly not sufficient condition for the success of this proposal. The complexities of inter-basin transfers, the jurisdictional arrangements at different levels (municipal and state) suggest that a new model of planning and management be established. A collaborative model with shared responsibilities must be sought.

The pioneering experience of legislation to protect water sources in the Metropolitan Region of São Paulo, carried out within EMPLASA (Metropolitan Planning Agency) during the 1970s, was an essential step for the integration between the water resources management and urban/metropolitan planning systems. Even if the water resources management system was not formalized then as it is today, the institutional communication and development of instruments intended to protect metropolitan sources was considered a great innovation. The alternatives to protect water supply sources that had been used until then involved either the expropriation of the areas to be protected, such as in the case of New York purchasing the land surrounding their main sources in New Jersey, or the adoption of structural measures (advanced treatment or inter-basin water transfers).

The main institutional factor faced by this innovative legislation related to water source protection was the total independence of the municipalities about land-use decisions. The independent municipalities showed little interest in cooperation with the rest of the region. However, this normative system brought together for the first time in Brazil a strategy for environmental protection with a strongly anticipatory character and moderate spending of public funds. It also launched the basis for intergovernmental cooperation on regulation of land use. These fundamentals were accepted by the new state legislation to protect sources (Law 9.866/97) that began to include principles of the water resources management system in its implementation strategy. In that law, the specific measures applicable to the water sources protected areas (WSPA) are defined in a decentralized form in the respective development and environmental protection plans (DEPP). These plans, articulated with the water resources management system, are the main instruments for decentralized land-use planning.

Most of the measures advocated by Law 9.866/97 are non-structural in nature, in the sense that they work with preventing pollution, not correcting it. Actions to discipline environmental quality, in the DEPP's are based on a basic strategy of interventions in basin areas such as: restricted occupation, oriented occupation and environmental restoration.

It is clear that it is possible for the managerial system of each WSPA to articulate its land use and occupation strategies directly with the appropriate municipal public powers. However, the outreach of such measures could be much better if articulated to a metropolitan planning system. It loses, first of all, due to the jurisdictional restrictions of the WSPA for not being planned in conjunction with other similar areas within the same metropolitan urban complex. Guidelines, for instance, concerning controls on occupation density are limited to the internal jurisdiction of the WSPA, which often may not be able to contain, in its territory, available area and infrastructure to accommodate growth pressures.

A second likely element of loss concerns the limitations of the sectorial jurisdiction as to the public functions that are of common interest. There are scales of these functions whose control is only defined as a metropolitan scope, and when this happens it will be very difficult to establish a communication between the strategic objectives of the DEPP and the function considered. This is the case, for instance, of communication with the urban transport systems. The only legitimate instance to establish a broad communication process between the public functions of common interest in the urban/regional complex as a whole is the metropolitan one, and this cannot be replaced in its sectorial scope by the water resources management system.

The need for integration between the water resources management systems and metropolitan planning results from acknowledging that the strict view on the local planning, when applied to urbanized basins, leads to irrational investments and sectorial system management. The management of sectorial systems, based on a predominantly local logic, creates the illusion that since they are different sectors converging on a restricted geographical unit, it promotes integration between sectors. However, this is not true if analysed in the light of functionality of sectorial systems as a whole (and not in segments).

Emerging Instruments for Total Urban Water Management

The Upper Tiete Basin Plan (FUSP, 2002) was developed according to objectives of an integrated view of management, in which water quality and quantity are dimensions that are understood to be common objects. This results in the need for effective coordination among the different water uses. In practice, this has proved to be a much more difficult task than it seemed to be when the general principles on multiple uses were first proposed. If quality and quantity are seen as functions of a same development and preservation policy, actions in the fields of source protection, sanitary sewerage (inside and outside the protected areas) and urban drainage, can no longer be treated as completely separate sectorial elements.

The integrating view proposed in the general objectives of the Plan challenges the underlying institutional organization of planning and management systems. On the one hand territorial communication involves different municipalities, or jurisdictions, and on the other hand there is the functional process, involving sectors that are based on specific planning, regulation, financing and operations. Both these integrating dimensions— territorial/jurisdictional and functional/sectorial—are equally important to achieve the proposed objectives. The institutional system for water resources planning and management will, in this sense, face four sets of challenges to integration.

- Integration between systems/activities directly related to water use in the river basin area, in particular water supply, wastewater treatment, flood control, irrigation, industrial use, energy use or other systems with a direct impact on the sources, such as that of solid waste, considering the perspective of a joint management of quality and quantity;
- Territorial/jurisdictional integration with instances of urban planning and management—the municipalities and the metropolitan planning system— considering the implementation of preventive measures for the urbanization

process, avoiding excessive demands on the quantities and qualities of existing resources, including flood events;

• Regulating communication with sectorial systems that are not direct users of the water resources, such as housing and urban transport, considering the creation of real alternatives to the process of occupying source protection areas and floodplains, and also making it possible to have urban development patterns that, as a whole, will not worsen the impervious conditions of the urban land and pollution throughout the basin;

• Communication with the neighbouring basins, in order to establish stable agreements on the current and future conditions of flows and export of water used in the basin. This type of communication will tend to become decisive and extremely complex in the future, given that all options to increase water offers to the MRSP, involve importing flows from other neighbouring basins;

All these are complex challenges and they cannot be fully met within the specific competencies of the water resources management system. The three former challenges require strong institutional communication with the environmental systems and metropolitan planning, as well as the relationship that may be established with the municipal governments involved, while the latter involves communication with the management systems of neighboring basins. Three major ensembles of actions are organized with a view to fulfilling the goals of the Upper Tiete Basin Plan (FUSP, 2002).

Outstanding in the first group of measures is the absorption by the basin management system of the large sectorial plans undertaken by the main agents such as SABESP (water supply and sanitation utility), the city of São Paulo, and the state water agency (DAEE). To reach an agreement between sectorial agents on such priorities of investments to implement structural measures is not easy, the reason being that sectorial agents are able to independently finance their own infrastrucutre needs. Sectorial priorities already defined were incorporated in the Bonn Plan, which allowed greater commitments from all actors involved.

Second, concerning institutional and legal measures, the outstanding aspect is the concept of flexible regulations. One of them is a system of stimulated adherence of the agents to the Plan goals, as a corollary of the recognition of its predominantly indicative character. Here, extremely innovative measures are considered, such as creating systems that will encourage state, municipal and private agents to guide their actions by the objectives of this plan, i.e. that will improve their performance with regard to the protection of sources and floodplain areas, management of water demand and rational use, management of solid wastes and groundwater management. For the water resources recovery and/or conservation goals that can be negotiated, there is a process of gradual compliance, in which the agent is encouraged to conform. These incentives may be expressed in financial benefits as well as by greater autonomy to define specific measures, with the decentralization of attributions of the basin management system in favor of the agent that complied with the main objectives proposed in the watershed plan. The flexibility thus induced, on the contrary of possibly weakening the management system, is a strategy that strengthens it by providing flexible instruments to review and re-structure specific procedures in order to fulfill its objectives effectively and efficient.

Third, measures concerning the improvement of the decision process, planning and management actions, constitute an essential counterpoint to the flexibility requirements assumed in the institutional strategy of the system. It considers the great dependence on flexible management for robust and transparent information systems to improve the quality of the decision. Key elements of the improved information system are the need for an increased surface water quality monitoring capacity, the development of a subsystem for information on groundwater and the integration of different sectorial bases on urban information.

Conservation and Rational Water Use

Another dimension of metropolitan planning, not covered by the water resources management system, is regulation. The regulating instruments, associated with the water resources management system, articulate water uses and services associated with them. Despite this, there are many aspects of the provision of services that indirectly interfere in the basin management strategy. One example is the coverage and equity in the basic sanitation services patterns within the urban sprawl. If there is no defined guideline among the municipalities that are part of the metropolitan region, regarding which areas should be the subject of priority actions, there is no way that the basin management system—*de motu proprio*, although in harmony with the sanitation services provider—can articulate its actions with this localized perspective of guiding urban development. It is not the objective of a water resources management system, and much less of the objective of the water supply utility or the drainage service or any other water use service, to establish overall strategies for urban/regional development. On the other hand, they are essential for more advanced communication of the basin management strategy with the water using service.

A broader view of the conservation and rational water use programs and actions is that it is needed as an essential element of an integrated concept of basin management systems and metropolitan planning. This type of program, largely based on demand management actions, is not always well understood with regard to its scope and extent.

The management of water demand in a basin with scarcity problems such as the upper Tiete is an essential measure to extend the relative offer, in the sense of increasing the number of users covered appropriately by the public supply system, maintaining the same water supply sources and flow. The water service operators are well informed concerning the specific programs to control losses in the network and on performing operational improvement to reduce loss of income. These actions are important and, in the case of reducing physical losses, imply effective gains of water availability in the basin. However, there are other actions to improve demand management that include reduced consumption, which as a rule are not included in the strategic planning of systems. Except for critical situations of scarcity and very high marginal costs to exploit new sources, momentarily below the maximum prices that can be practiced in the tariff system, it is not the job of the water service to promote stable programs for management of demand. In this context the PURA Program of SABESP is a positive exception due to the fact that it supports initiatives to improve constructive systems and sanitary education, which may lead to reducing demand. But it is not reasonable, in a strategic concept of demand management in the basin, to concentrate all of the responsibility of demand management within the scope of the provider of the supply service.

A stable program for demand management, in the terms of the strategic principles of the Upper Tiete Basin Plan, requires a decision and managerial structure independent from the water supply company, although nothing will prevent the latter from continuing to hold the executive system and playing a central role in defining the specific priorities.

The range of appropriate activities to establish a steady policy for conservation and rational water use involves, besides measures directly related to the reduction of losses and abusive uses, several possibilities of interaction with the municipalities and the metropolitan planning system. Measures such as the limited use of 'water brooms' and washing vehicles, or the adoption of good practices to design and construct buildings systems, including the subsidized exchange of water consuming equipment, are only implemented with the help of municipal government and through a metropolitan policy directed specifically to these objectives.

The structure of a metropolitan program for water conservation and rational use must be based on active managerial competence, which can clearly measure the costs and benefits associated with each of the levels of action advocated. Thus, different levels of advance and complexity of demand management actions may be rationally defined and compared with the costs and benefits of increasing the offer. This strategic vision of management of demand is not only applied to exploring new sources (in the sense of ensuring that the cost of recovered flow is not greater than the new source to be exploited) but also with respect to the operational regimes of each reservoir considering the hydrologic risks of failure. The institutional competency to implement a program with these characteristics appears to accrue naturally to the basin management system, more specifically to the Basin Agency. However, the communications with broader urban and housing actions once again lead to the need for communicating with an active metropolitan agency.

Conclusions

It is clear from the points raised in the previous sections that proper water management in megacities requires an integrated approach. The traditional approach for dealing with urban water problems is fragmented. There is a body of technical knowledge and an institutional framework to deal with water supply and sanitation. Another one deals with urban drainage, another deals with solid waste disposal, another one deals with housing, and finally, another one deals with transportation. In reality all these sectors interact from both a technical and institutional point of view.

Water management of megacities in the developing world is probably the major environmental challenge of the future. As the urban population increases the problem tends to be magnified. Fragmented management in which the sectors involved (water supply and sanitation, drainage, solid waste, housing and transportation) are dealt with separately from both the technical and institutional viewpoints, and this is observed all over the world. This approach has been responsible in part for escalating degradation of the urban ecosystem.

Metropolitan São Paulo is an excellent case study of this complex situation. The population has increased fifty-fold in the last century. City water managers face a tremendous challenge of providing water to a population of 18 million. Solid waste generation is making urban drainage management more complicated. Lack of wastewater treatment is another threat when summer flooding occurs. Fortunately, some signs of improvement are being observed in changing this situation. The paper has shown that the solutions to the problems involve huge investments in infrastructure for water supply, sewage treatment and flood control. At the same time an integrated management approach, TUWM, will allow planning of the correct infrastructure and its operation in a complex management setting.

TUWM brings together all these sectors under a common umbrella. This does not mean that a single major institution is responsible for all the actions. The complexity of the problem suggests that the river basin committee is not plural enough for decision-making at this level. Hence, it is proposed that a new metropolitan board including the stakeholders (public and private) of the different sectors mentioned is formed. This board would be responsible for proposing integrated actions and ensuring that public funds are better allocated. It is important that the board reports to the highest political authority of the region. In the MRSP this would be the Governor of the State of São Paulo. Total urban water management maybe a way to move from the present chaotic situation to livable cities in the future.

Note

1. A lot is a parcel of land in the urban space.

References

Braga, B. P. F. (2000) The management of urban water conflicts in the metropolitan region of São Paulo, *Water International*, 25(2), pp. 1–6.

Canholi, A. (2003) The flood control project of the Aricanduva River Basin, *Proceedings of the XV Brazilian Water Resources Symposium, ABRH*, pp. 45–49, (Curitiba: PR) (in Portuguese).

Fundação de Apoio à Universidade de São Paulo (FUSP) (2002) Upper Tiete River Basin Plan. Final Report, São Paulo (in Portuguese).

SABESP (2004) Potable Water Supply in the Metropolitan Region of São Paulo: Present and Future, *Cia. de Saneamento de São Paulo (SABESP)*, (São Paulo).

Water Management in Mexico City Metropolitan Area

CECILIA TORTAJADA

Introduction

Mexico is a country of contrasts and disparities from economic, social, environmental and cultural viewpoints. Over time, economic and social inequities have become increasingly more acute among the regions. While there are people with access to education, health, water, electricity, roads, infrastructural services, etc., there are many others who lack access even to the most basic services.

In terms of water resources, there is an enormous imbalance between water availability and its use. The main economic activities are concentrated in the central, northern and north–western regions of the country, representing approximately 84% of the GDP, but with a per capita water availability of only 2044 m^3/year. On the other hand, in the south–eastern part of the country, where water availability is 14 291 m^3/year/per capita, only 16% of the GDP is produced. This region has the highest rates of poverty in the country and lacks most types of infrastructural development (OECD, 2002a, 2003; CNA, 2005). This means that in the regions endowed with more natural resources, including water, poverty is more acute due to an unfortunate combination of lack of appropriate policies and institutions, which, among other issues, have negatively affected the quality of life of the local populations and the environment in which they live.

At the beginning of the 20th century, approximately 80% of the population in Mexico lived in settlements of less than 2500 people. However, by 2000 60% of the population lived in settlements with more than 15 000 people. The increase in concentration of population in urban and peri-urban areas, many of them living under conditions of extreme poverty, has resulted in increased pollution and other stresses on water resources and infrastructure.

Approximately 30 million people currently live in settlements with less than 2500 people. These heavily marginalized areas have low economic productivity, high unemployment and outmigration rates, and poor access to services like education, health, clean water and sanitation. Malnutrition, low life expectancy and high mortality rate are also the highest in these areas (OECD, 2002b).

In 2005, in terms of water-related services, 89.8% of the population at the national level had access to drinking water and 77.6% to sewerage. However, more than 11 million people still lack access to drinking water and over 22 million do not have access to adequate sanitation, with the rural areas generally lagging behind in terms of having adequate services (Gobierno Federal, 2005). Coverage of water supply, in the present context, refers to the population that have access to piped water in their houses or their properties, and to the population able to obtain water from other houses, properties or from a public source. In terms of sanitation, the Mexican statistics include population connected to a public sewer and septic tank, and those discharging wastes directly to rivers, lakes or ravines. Data are not available on either quality or reliability of the services received.

In the case of the capital city, the increased urbanization and high population growth within Mexico City and the neighbouring State of Mexico resulted in the designation of an area known as Mexico City Metropolitan Area (ZMCM, by its acronym in Spanish). This metropolitan area, with an approximate population of 20 million inhabitants and industries, services and commercial activities that generate 33.2% of the GDP, plays very important roles in the country, both from economic and political viewpoints (SEMARNAT/CNA, 2000). However, it faces escalating demands for services in areas like water, sanitation, electricity, education and health, among many others.

Mexico City Metropolitan Area

Mexico City is the capital of Mexico. It is located in the Federal District at 2240 m above the mean sea level (msl) in the south–western part of the Valley of Mexico. It is surrounded by mountains reaching a height of over 5000 m above msl.

At the beginning of the 20th century, Mexico City was still in the north-central area of the Federal District. However, due to increased urbanization, its 16 boroughs at present cover its entire surface area. In fact, according to the Mexican Constitution, at present Mexico City is equal to the Federal District, and both terms refer to the same location.

At present, approximately 9 million people live in 60 203 ha of urban areas and 88 442 ha of rural or conservation areas (land that is left in its natural state, often for groundwater recharge) (see Figure 1). However, these figures do not represent the reality since both rural and urban development have taken over a great part of the conservation areas (PNUMA et al., 2003).

The Federal Government, and much of the industries, education and employment facilities and cultural centres of the country are concentrated in Mexico City. However

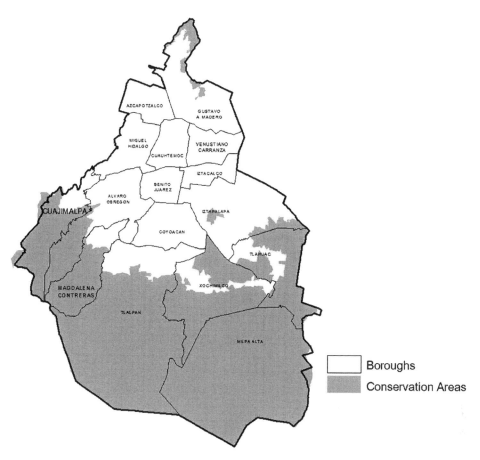

Figure 1. Conservation areas in the federal district. *Source:* Centro de Investigación en Geografía y Geomática 'Ing. Jorge L. Tamayo', Mexico.

the quality of life for the population has decreased significantly in recent years, primarily because of increasing population density, and extensive air, noise and water pollution.

With regard to the ZMCM, in 1990 it included the 16 boroughs of Mexico City and 27 municipalities of the neighbouring State of Mexico. In 1995, it was decided to include within the ZMCM the municipalities of the State of Mexico, which had the highest population as well as economic growth. At present, according to the National Council for Population (CONAPO, 2000), the metropolitan area includes the 16 boroughs mentioned above, 37 municipalities of the State of Mexico and one municipality of the neighbouring state of Hidalgo. However, according to the National Institute of Statistics, Geography and Informatics (INEGI), the ZMCM includes the 16 boroughs of Mexico City and 34 municipalities of the State of Mexico.

The ZMCM covers an area of 4925 km^2 (1484 km^2 in Mexico City, and 3441 km^2 in the State of Mexico), representing about 0.25% of the national area. The population density varies from 13 500 to 131 persons/km^2. The State of Mexico is the most populated area in the country, with 13.1 million inhabitants, followed by Mexico City, with 8.6 million

(INEGI, 2000a). The State of Mexico also has the highest population growth rate of all the states in the country, including Mexico City. During the period 1990–2000, this state had an annual population growth rate of 2.9%, whereas the Federal District had an annual growth rate of 0.4%.

Urban growth in the ZMCM has been very rapid and disorganized, which has resulted in acute environmental deterioration, including water and air quality. The rapidly increasing urban settlements continue invading what used to be protected land, and land use has changed from forestry to agricultural, and finally to urban. This uncontrolled growth in the ZMCM has progressed towards both the State of Mexico and to the rural areas of Mexico City (PNUMA *et al.*, 2003).

The expanding population, as well as the rapidly increasing industrial, services and commercial activities, have represented a formidable challenge for the institutions responsible for providing the necessary services, including water and sanitation, primarily in terms of management, investments and energy consumption. The investments have not only represented high economic costs but also high social and environmental impacts, which have become almost unmanageable (INEGI, 2001).

Historical Development of Mexico City Metropolitan Area[1]

The water supply and wastewater systems in the metropolitan area, as well as in any other locations, cannot be analysed without considering the associated human and geographical environment. They have to be considered in relation to issues such as geography, climate, population growth, urbanization, migration, economic development and social expectations. In the case of the ZMCM, the evolution of the management of water and wastewater systems should be seen as an integral component of a rapidly expanding metropolitan area. Therefore, an overview of the changes that have occurred in the metropolitan area during the last 65 years will provide a better understanding of the water supply and sanitation situation in the region.

In 1940, the Federal District had a population of 1.75 million people, of which 1.6 million lived in the downtown area (what was then known as Mexico City). During this decade, the metropolitan area started to grow mainly as a result of increasing economic activities in the municipalities adjacent to the Federal District.

In 1950, the metropolitan area included Mexico City as it was then, seven boroughs of the Federal District and two municipalities of the State of Mexico. The population was 2.9 million people, living in an urban area of 26 275 ha, with a population density of 113.5 people/ha. Population density was higher in the downtown area compared to the rest of the boroughs, which included mainly rural settlements (less than 2500 people).

During this decade, Mexico City developed primarily towards the north, reaching the limits of the State of Mexico. This resulted in increasing urban activities on both sides of the border, and industrial activities primarily were in this state. The National Autonomous University of Mexico was established in the southern part of the City. This was followed by progressive urban development in this area, with middle and high-income settlements as well as industrial activities. During this period, the government of Mexico City decided not to authorize any additional housing construction. This resulted in formal and informal urban developments in the State of Mexico.

In 1960, the metropolitan area at that time included Mexico City, 15 boroughs of the Federal District and four municipalities in the State of Mexico. The population had increased to 5.1 million inhabitants within an urban area of 41 690 ha, which resulted in a population density of 123.66 persons/ha. This was an increase of almost 73% in terms of population, and more than 58% in urban area, compared with the situation in 1950.

During this decade, Mexico City, as well as the metropolitan area, changed dramatically not only due to population growth, but also due to very rapid urban, road and industrial developments. There was an huge expansion of planned high-rise buildings for low and medium-income families, as well as unplanned settlements. Restrictions for the construction of housing continued in the Federal District, which resulted in an increasing number of informal settlements in the City.

In 1970, the metropolitan area included Mexico City, 16 boroughs of the Federal District and 11 municipalities of the State of Mexico. The population had increased to 8.6 million inhabitants and the urban area had reached 72 246 ha. The urban land used increased by 73% and seven municipalities were added to the metropolitan area, which reduced the population density to 120 persons/ha. This period witnessed a massive urban expansion of both formal and informal settlements within the overall ZMCM.

In 1980, the population in the metropolitan area had increased to 13.7 million (59% compared to 1970) and the urban area by another 89 112 ha (23%) compared to 1970. The population density had increased to 154 persons/ha.

Between 1980 and 1990, the population in the metropolitan area increased to 15 million people and the urban area covered a total of 40 390 ha (11 306 in the Federal District and 29 084 ha in the State of Mexico), with the highest urban growth in the State of Mexico.

From 1950 to 1995, the population of Mexico City increased from 3 to 17 million people (Table 1). In contrast to the previous decades, during the period between 1990–2000, the annual population growth of Mexico City was only 0.4%, compared to the ZMCM, which was 2.9%. The main reason for the growth in ZMCM was immigration from the rural areas and from the medium and small-size cities.

Throughout these decades, the population growth and the planned and unplanned urbanization have resulted in an uneven race for the Federal and the local governments to construct an infrastructure and provide essential services to the population, including water supply and sanitation. An example is the number of houses, which almost tripled in 40 years, with more than 1.7 million in 1990 compared to approximately 600 000 in 1950.

Demographic and socio-economic conditions have had a major influence on the overall urban growth, and land use has depended on the social and economic conditions of the local population. At present, about 67% of the population in the ZMCM can be considered to be at the medium to low socio-economic level, about 15% are in high and medium to high, and 18% in the very low (PNUMA *et al.*, 2003). This means that the land use of the different parts of the metropolitan area reflect the needs and the opportunities of their

Table 1. Average population of Mexico City Metropolitan Area, 1950–95 (millions)

	1950	1960	1970	1980	1990	1995
ZMCM	2 982 075	5 155 327	8 656 851	13 734 654	15 047 685	16 898 316
Mexico City	2 923 194	4 846 497	6 874 165	8 831 079	8 235 744	8 489 007

Source: CONAPO (2000).

population. The wealthier areas have better overall living conditions and more assured access to services (e.g. access to water supply and sewerage, collection of solid wastes, schools, hospitals, road infrastructure, etc.) compared to the less wealthy areas.

Table 2 shows the access in the ZMCM to services such as electricity, water supply and sewerage. More people in Mexico City now have access to such services, compared to the municipalities in the State of Mexico that are part of the ZMCM. This is because Mexico City, being the capital of the country, is much more urbanized than the municipalities, and it has also more economic and political power (Tortajada, 2006).

A large number of high-income houses in Mexico City are not connected to the public sewer because they have been constructed on volcanic rocks, which has made it difficult and expensive to build such an infrastructure. This is especially the case for many settlements in the southern part of the City, most of which have septic tanks that are frequently not properly constructed and managed.

For a region with a population of more than 20 million people, which is steadily increasing, the provision of all services, including water supply and sewerage, has been a challenging task. The responsibilities for water supply and sanitation have been exclusively in the hands of the different levels of governments, whose lack of planning, managerial and human constraints, and political interference, have been obvious throughout the years. In addition, the population in general has not developed any sense of responsibility or interest in participating in the conservation, protection and management of water resources, even though, in the final analysis, water supply and sanitation is for their own benefit and use.

Urban Growth and Conservation Areas

Conservation of the rural areas is fundamental for the water security of Mexico City since it has a direct bearing on groundwater recharge. Rural areas within Mexico City are considered to be conservation areas under the Law of Urban Development.[2] In Article 30.II, the Law defines a conservation area as:

> the land which should be considered as such according to its location, extension, vulnerability and quality; that which has an impact on the environment and on land-use planning; mountains and areas useful for the recharge of the aquifer; hills, valleys and elevations which are natural elements of the land of the City; and land for agricultural and livestock activities, for fisheries, forestry, agroindustry and tourism, as well as rural settlements.

Conservation areas are increasingly threatened because of steady urban growth. Between 1980–2000, 76% (377 000 units) of the new houses that were constructed in Mexico City were located in the seven boroughs with the most conservation areas. Of the 44 rural settlements that still exist in Mexico City, 35 of them (400 000 people) are located in conservation areas.

Expansion of illegal settlements has also become a critical problem for the City. In 2003, there were 804 so-called 'irregular settlements' with approximately 60 000 families, living in 2400 ha of land for periods of between 10 and 22 years (SMA Programa de Protección Ambiental del DF 2002–2006, in PNUMA et al., 2003). Some 80% of these

Table 2. Access to services in the ZMCM in urban AGEB[a], 1990

	Total	Downtown	Municipalities in State of Mexico within the Metropolitan Area	Low-income settlements	High-rise buildings	Middle-income settlements	High-income settlements	Other
Number of houses without electricity	54 048	788	6814	40 461	3175	3598	704	329
Percentage of houses without electricity	1.8%	1.6%	2.7%	2.1%	0.7%	0.9%	1.1%	1.7%
Number of houses without sewerage	545 836	2247	96 010	398 218	23 707	17 122	8532	3249
Percentage of houses without sewerage	17.6%	4.5%	38.5%	21.1%	5.1%	4.4%	13.5%	17.2%
Number of houses without tap water	1 115 262	6486	133 878	872 222	61 026	36 341	5309	5932
Percentage of houses without tap water	35.9%	12.9%	53.5%	46.2%	13.2%	9.3%	7.8%	31.5%
Number of private houses	2 147 341	24 075	189 214	1 256 228	375 017	257 919	44 888	9996
Number of rented houses	678 956	20 837	40 903	448 443	53 788	104 927	10 058	5444
Percentage of private houses	69.2%	48.0%	75.7%	66.5%	81.2%	65.8%	76.0%	53.1%
Percentage of rented houses	21.9%	41.6	16.4%	23.7%	11.6%	26.7%	17.1%	28.9%

[a] Urban AGEB refers to geographical areas in settlements consisting of 2500 people or more (all municipalities are included even if population is less than 2500). Land use is for housing, industries, commercial, recreation or any other use, but not for agriculture, livestock or forest.
Source: Resultados definitivos. INEGI, Datos por AGEB Urbana, XI Censo General de Población y Vivienda 1990. Volúmenes del Distrito Federal, Estado de México e Hidalgo, 1992, in: CONAPO (2000).

families are in conservation land, and about 20% of them live in dangerous places such as river beds.

Even though there are urban land-use programmes whose main objective is to control the expansion of rural and irregular settlements in conservation areas, the demand for all types of settlements has been overwhelming. It has simply surpassed any attempts by the public institutions to catch up with the demands for housing and infrastructure, and to provide appropriate services. The net result has been that people often do not have access to even basic services such as electricity, water supply and sewerage, which is especially relevant for settlements in conservation areas. Overall, the demands for housing and infrastructure by all socio–economic levels have continually increased. The unsustainable urban growth and inadequate management have resulted in a mounting pressure on the Federal and local governments to provide more and better services. At the same time, people are now reluctant to live in a polluted and unsafe environment (PNUMA *et al.*, 2003).

Water Availability

The water supply in the ZMCM depends primarily on local groundwater sources and on interbasin transfers. Mexico City, and the most populated 17 municipalities of the State of Mexico, share the same sources of water, as well as the infrastructure for water distribution.

In 2002, the volume of water supplied to the ZMCM was 2.236 MCM/day (1.200 MCM/day from 374 deep wells; 0.071 MCM/day from 18 springs only for Mexico City, and 0.964 MCM/day from 97 sources of water, such as snowmelt in the case of State of Mexico[3]) (INEGI, 2003). The second main source of water is the Lerma-Balsas and the Cutzamala River systems, which will be discussed later. It is estimated that the ZMCM receives $66 \, m^3/sec$ mainly for domestic supply, with Mexico City receiving about $35 m^3/sec$ and $31 m^3/sec$ for the State of Mexico (see Table 3). Within Mexico City, the water is distributed to the users through a primary network of 1074 km of pipelines (with diameters of 0.5–1.83 m) and a secondary network of 12,278 km (with diameters of less than 0.50 m). The water supply system comprises 16 dams having a total storage capacity of $2827.90 \, km^3$ (INEGI, 2000b). Data are not available for the municipalities in the State of Mexico.

Table 3. Water supply sources for Mexico City Metropolitan Area

	Federal District (m^3/s)	State of Mexico[1] (m^3/s)	Total (m^3/s)	Percentage
Internal sources	20.0	25.2	45.2	68.5
Wells	19.0	24.8	43.8	66.4
Springs and rivers	1.0	0.4	1.4	2.1
External sources	14.8	6.0	20.8	31.5
Cutzamala	9.9	5.0	14.9	22.6
Lerma	4.9	1.0	5.9	8.9
Total	34.8	31.2	66.0	100.0
Percentage	52.7	47.3	100.0	

[1] Only municipalities which are part of the ZMCM.
Source: INEGI (2003).

In 2000, 95.3% of the population in Mexico City and 84.2% in the State of Mexico had access to water, either with a water connection directly to the house or from common faucets in the neighbourhood (INEGI, 2000a). However, most of the aquifers, springs and rivers which supply water to the ZMCM are located to the west, north and south. Thus, water supply is irregular and unreliable for the people living in the eastern part, who are also most affected by water shortages.

More than 5% of the people living in the metropolitan area still do not have access to water. While some of them receive water from the government in pipes, people have to pay water from private vendors. The cost of water (200 litre-containers) often represents between 6 to 25% of their daily salaries. Poor people who buy water from trucks pay around 500% times more than the domestic consumers. In addition, drinking water for much of the population in the ZMCM comes from 20–30 litre-containers of purified water, which are sold commercially. The reason for this is the near universal distrust of the quality of the tap water. This means that not only people with no access to tap water spend a certain percentage of their income buying bottled water, but also people with access to tap water have to buy containers of water, of which the quality control leaves much to be desired. Indeed, Mexico is the second largest consumer of bottled water in the world, with a consumption of 168.6 l/person in 2004 (Rodwan, 2004).

Main Sources of Water for Mexico City Metropolitan Area

The ZMCM is located in the Valley of Mexico basin, which is surrounded by the basins of Lerma, Cutzamala, Amacuzac, Libres Oriental and Tecolutla rivers (INEGI, 2001). The Lerma and the Cutzamala river basins, together with the aquifer of the Valley of Mexico, are the main sources of water for the metropolitan area. The aquifer of the Valley of Mexico contributes 70%, the Lerma-Balsas river basin 9% and the Cutzamala river basins 21%. The very few surface water bodies that still exist in the basin of the Valley of Mexico provide only 2.5% of water supplied (CNA, 1997a; INEGI, 2000b).

Mexico Valley Aquifer

The annual rate of withdrawal from the aquifers is significantly higher than the recharge rate: $45–54 \, m^3/s$ is abstracted each year, but natural recharge rate is only about $20 \, m^3/s$. This mismatch has resulted in a significant over-exploitation, which has contributed to the lowering of the groundwater table by about 1 m per year. Steady lowering of the groundwater level increased the land subsidence rate, initially to 10 cm/year, and later up to 30 to 40 cm/year. The average annual subsidence rate in the area of the International Airport of Mexico City is 20–25 cm, and in the City Centre it is around 10 cm. It is estimated that the central area of the metropolitan area has subsided by 10 m during the past 100 years (Legorreta *et al.*, 1997; INEGI 2001; Gobierno del Distrito Federal *et al.*, 2004).

However, the problems related to water supply in the metropolitan area extend well beyond the subsidence of the City. For example, the water supply and drainage systems have not only become very large and complex, but also obsolete in many areas. Provision of water services varies in the different parts of the City, tariffs are still heavily subsidized, the quality of water supplied is poor, levels of unaccounted for water are unacceptably high and the population wastes enormous amounts of water. People living in wealthy areas use up to 600 litres per capita per day, while the corresponding rate in the poor areas is about 20 litres.

Use of deep wells has resulted in an increase in the iron and manganese content of the water, thus decreasing water quality and making water treatment more expensive. Water infrastructure has become more vulnerable to earthquakes. Over-exploitation is reducing soil moisture in the surrounding mountains, which is damaging forest cover and adversely affecting the ecosystems.

A very high percentage of water is lost from the distribution networks because of leakages and illegal connections. Inappropriate overall management, aged pipes, inadequate maintenance over prolonged periods, poor construction practices and continuing land subsidence, are contributing to high levels of unaccounted for water. It is estimated that more than 40% of water is lost in the network due to leakages, which represents about 130 l/person/day. It is estimated that this volume of water would be enough to provide a service to 4 million people (UNAM, 1997; Secretaría de Obras y Servicios, 2002).

Lerma Valley Aquifer

In 1942, the Lerma Valley project (62 km from Mexico City) was initiated to meet the steadily increasing demands of water in the metropolitan area. The first stage was planned and constructed to bring 4 m^3/sec of water to the metropolitan area. It included the construction of five wells between 50 and 308 m deep for groundwater abstraction, and a 62 km, 2.5 diameter pipe for its distribution. This pipe is laid along the *Sierra de las Cruces*, through the 14 km long *Atarasquillo-Dos Rios* tunnel. Four tanks, 100 m in diameter, and 10 m in depth, were built in Mexico City for storage. This water is then distributed to the City by gravity. The increasing demands for water resulted in the construction of the second stage of the project. Between 1965–75, some 230 deep wells were dug, which increased the volume of water abstracted to 14 m^3/sec. However, due to environmental impacts and social conflicts, the volume abstracted later had to be reduced to 6 m^3/sec (Legorreta *et al.*, 1997).

The political relationship between the authorities of Mexico City and the State of Mexico have been strongly influenced by the social conflicts that have resulted from the interbasin transfer of water from the Lerma Valley to the metropolitan area. The main interest of the Federal and the Mexico City governments has been primarily to guarantee water supply to Mexico City. As a way of compensating the local populations, small projects were constructed in the towns that were adversely affected by the water transfer project. The over-exploitation of the aquifers in the Lerma area has reduced the fertility of the soils. Agriculture has now become mainly rain-fed, and not irrigated as previously. The economy of the region and the life of the population have changed significantly (Legorreta *et al.*, 1997).

Cutzamala System

In 1976, the 'Cutzamala System' was planned to supply water to the metropolitan area from the Cutzamala river, and thus reduce the over-exploitation of the Mexico Valley aquifer. The water is transferred from a distance of 60 to 154 km, and then pumped to a height of more than 1000 m, requiring 102 pumping stations, 17 tunnels and 7.5 km of canals, which makes this project extremely energy-intensive and expensive (CNA, 1997a)

Initially, what later became the Cutzamala System, was planned as a hydropower project, called the Miguel Aleman Hydroelectric System. Cutzamala started by taking advantage of the infrastructures for hydropower generation, but the planned water use was changed. Currently, only 3 m³/s is used to generate hydropower during peak hours and to satisfy the local energy requirements for the agricultural and industrial sectors (CNA, 1997a). Due to the magnitude of the project, its construction was initially planned in three stages. The first stage has been in operation since 1982 (4 m³/s), the second since 1985 (6 m³/s) and the third one since 1993 (9 m³/s) (CNA, undated, a). During the first stage of the project, water was brought from Victoria Dam and was distributed through a 77 km long and 2.5 m diameter aqueduct, which crosses the *Sierra de las Cruces*. The second and third stages of the project included the construction of both a water treatment plant and a central aqueduct. The implementation of these two stages was very complex mainly due to the height to which the water had to be pumped: 1100 m. The electricity used to pump the total volume of water from the Cutzamala system just to the treatment plant is the equivalent of the energy that is consumed by the city of Puebla, with a population of 8.3 million people (Legorreta *et al.*, 1997). An overview of the infrastructure for Cutzamala System is presented in Figure 2. The elevation at which the different dams and pumping plants of the System are constructed is included.

In terms of investments, according to the EIA carried out for the fourth stage of Cutzalama, the total cost of the first three stages of Cutzamala was $965 million (1996 estimates). If the estimated cost of the earlier hydroelectric plant is added, the total investment cost becomes $1300 million. The cost of the cancelled hydropower system, with a total installed capacity of 372 MW, has been estimated at $325 million, at an average cost of $875 000/MW. The total cost of the Cutzamala System at $1300 million (mainly construction and equipment costs) was higher than the national investment in the entire public sector in Mexico, in 1996, in the areas of education ($700 million), health and social security ($400 million), agriculture, livestock and rural development ($105 million), tourism ($50 million), and marine sector ($60 million). Up to 1994, the Cutzamala System alone represented three times the annual infrastructure expenditure of the Ministry of Environment, Natural Resources and Fisheries for 1996, which was more than $470 million (CNA, 1997a).

The annual energy requirements necessary to operate the Cutzamala System are about 1787 million kWh, representing an approximate cost of $62.54 million. The investment would increase significantly if the costs in personnel ($1.5 million/year) and water treatment process costs were added (CNA, 1997a). If only the operational costs for running the Cutzamala System are considered (about $128.5 million/year), supplying 600 million m³ of water (19 m³/s) would mean an average cost per cubic metre of water of $0.214 and an energy consumption of 6.05 kWh/m³. Hence, the price charged to the consumers, about $0.2/m³, is not enough to cover either the operational costs of the Cutzamala System or the treatment and distribution costs of water to the metropolitan area.

In addition to the construction of the Cutzamala, about 190 so-called 'social projects' were built for the benefit of some of the people living in the municipalities who are mostly affected by water shortages. These projects were built jointly by the National Water Commission of Mexico (CNA) and the communities, and consist mainly of construction, enlargement and rehabilitation of water supply and sanitation systems, as well as construction and rehabilitation of houses, schools and farms. Equally important was the construction and the rehabilitation of roads by CNA, both for Cutzamala and the local

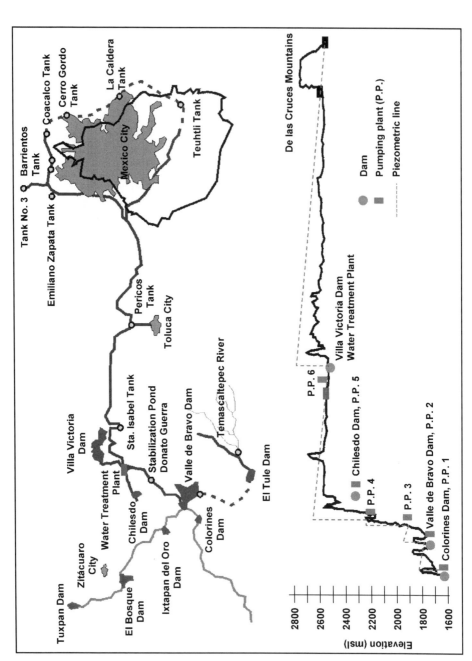

Figure 2. Overview of the infrastructure for Cutzamala System. *Source:* IMTA (1987).

population. In 1996 the cost of these 'social projects' was estimated to be the equivalent of 5% of the direct investment of the Cutzamala, which represented an additional $45 million (CNA, 1997a). A very important issue that has not been resolved as of March 2006, has been the resettlement of the affected communities due to the construction of the Cutzamala project, who after all these years, still have not received the expected compensation.

The programme on drinking water and sanitation of the metropolitan area considered the construction of a fourth stage of Cutzamala which would increase the volume of water transferred to the Valley of Mexico from $0.6\,km^3$/year ($19\,m^3$/s) to $0.76\,km^3$/year ($24\,m^3$/s), and the treatment of $1.3\,km^3$/year ($42\,m^3$/s) of wastewater. The fourth stage of Cutzamala (Temascaltepec project) was to be initiated in 1997. This stage included the construction of a 120 m high dam, 743 m in length at the crest. The reservoir would have a capacity of 65 millions m^3 and regulate an average flow of 5000 l/s. The project envisaged a $15\,m^3$/s pumping station and the construction of 18 km of canals and 12 km of tunnels (CNA, 1997a). The water would flow to the Valle de Bravo Dam through a 18.75 km long and 3.5 m diameter tunnel. According to official figures, the initial investment was estimated to be $502 million. Once the fourth stage of the Cutzamala is operational, the volume of water will increase only by $5\,m^3$/sec of water, from 19 to $24\,m^3$/sec (Tortajada, 2001).

As of March 2006, the Temascaltepec project has not yet been started because of serious social constraints. The population of some of the villages of Temascaltepec are afraid that the construction of the tunnel will dry up springs (El Naranjo, La Huerta, El Sombrero y El Chilar) and will affect the agricultural production of the area (maize, sugar cane, banana, tomato, melon and peas). Even though the local people who would be affected by the project are against the project (Legorreta *et al.*, 1997; Agua Latinoamérica, 2004; La Jornada, 15 July 2004; El Universal, December 2005), authorities consider the development of Temascaltepec river to be of the utmost importance for the development of not only Mexico City, but also of the State of Mexico as noted in the Development Plan of the State of Mexico 1999–2005 (Government of State of Mexico, 1995).

For years, studies have indicated that if the leakages in the distribution system in the ZMCM were repaired, there would be no need to construct the fourth stage of the project. This means that the additional water supply of $5\,m^3$/s that is being planned with very high economic, social and environmental costs would not be necessary. However, this type of rational planning and management continues to be absent in the relevant water management institutions.

In addition to Cutzamala, the other sources of water that the Federal Government has identified for potential contributions to the water supply of the metropolitan area are the Amacuzac, Tecolutla and Atoyac rivers (Gobierno del Distrito Federal *et al.*, 2004). The Amacuzac river project would include the construction of a 185 m high and 450 m wide dam, with an inundated area of $67\,km^2$, and a storage capacity of 4000 MCM. The dam would be located in the borders between the states of Morelos, Guerrero and Puebla. Water distribution from this site to the ZMCM would require the construction of a 160 km long aqueduct, and, depending on the final design, either two pipes of 4.5 m of diameter or three pipes of 3.5 m diameter. Water would have to be pumped to a height of 1825 m, requiring a generating capacity of 4000 MW. The annual electric power consumption for this system is estimated to be 5% of the annual national electric power production, representing 16.5 million barrels of oil per year. It is claimed that this project will make it unnecessary to abstract $50\,m^3$/sec of groundwater from the Valley of Mexico aquifer any more. The rational is that the groundwater would be used only during periods of severe droughts or when the other water distribution systems were not working due to maintenance activities (Tortajada, 2004; CCE and CMIC, 2000).

Under these conditions, it will certainly be more economical, socially acceptable and environmentally desirable to consider first demand management practices such as a reduction in the unaccounted for losses, water pricing and other water conservation practices, before embarking upon extremely expensive new water development projects, with high social and environmental costs. It has been estimated that each cubic metre of water from the Cutzalama river required an investment of $23 million. This estimate would increase by a factor of four if the source of water were the Amacuzac river (INEGI, 2001).

Basically, governmental institutions in the past have ignored the potential social conflicts and disruptions that could result from interbasin water transfers. In addition, no authoritative analyses have been made on the nature of the beneficiaries and the people who may have to pay the costs. Surprisingly, even the EIA for the fourth stage of the Cutzamala System (CNA, 1997a) does not consider any social costs. As with most of the EIA that are carried out in Mexico, mostly physical technical factors are considered: social issues are conspicuous by their absence (Tortajada, 1999, 2001). In 2003, the government of the State of Mexico took the government of Mexico City to court and demanded compensation of $2.2 billion due to damage caused by over-exploitation of the aquifers and excessive abstraction of water to the detriment of people in the State of Mexico. The decision of the Supreme Court was expected to set precedents for similar cases in the future. However, in October 2005, the newly elected Governor of the State of Mexico publicly declared that he would withdraw the court case, since he preferred to work with the Federal and Mexico City governments to find an amicable solution.

Aqueducts: 'Cutzamala-Macrocircuito' and 'Cutzamala-Aquaférico'

The Federal Government, as well as the government of the State of Mexico and CNA, initiated the construction of two distribution lines in 1980 to ensure a more efficient distribution of water from the Cutzamala System. Mexico City was constructing a water distribution system known as *Aquaférico* which would come from the west, and would supply water to the southern and eastern parts of ZMCM.

In the State of Mexico, the water distribution system is known as *Macrocircuito*. The construction was planned around most of Mexico City towards the north, carrying water to the northern, southern and eastern parts of the City (CNA, undated, a). The first stage of this system was inaugurated in October 1994. Both the first and the second stages are now in operation and provide a continuous supply of $4 \, m^3/s$. This has benefited around 1.4 million people, with a supply of 250 l/capita/day. The operation of the third and fourth stages planned to increase water availability by an additional $7 \, m^3/s$ (total volume of $11 \, m^3/s$), benefiting 4 752 000 inhabitants who live in the eastern and northern parts of the State of Mexico, with approximately 200 l/day/person (CNA, undated, b, c, d; CNA, 1997b). The system includes the construction of two pipelines, with a total length of 168.28 km. This is in addition to 58.28 km of pipelines that have already been constructed. The two pipelines will require a surface area of 336.56 ha, plus 71 ha for the storage tanks (CNA, 1997b).

The total investment costs for *Macrocircuito* between 1987 and 1997 were $78 million, while the estimated cost for the third and fourth stages (1997–2000) were expected to be approximately $190 million, making a total investment of $268 million. This amount represents almost half of the total public sector budget at the national level for 1995 ($563 million) in the areas of urban development, ecology and drinking water (CNA, 1997b).

The projects were expected to be completed by 2000, but so far the construction has progressed very slowly (CAEM, 2003; Reforma, 9 November 2004).

Wastewater Management

The soil of Mexico City is basically clay, and thus susceptible to compaction. Accordingly, the higher the volume of water abstracted, the higher is the rate of land subsidence (CNA, 1997a). The sinking of the City has resulted in extensive damage to its infrastructure, including the water supply and sewerage systems and degradation of the groundwater quality. It has also required the construction of costly pumping stations to remove wastewater and stormwater from the City.

At the beginning of the last century, the sewerage system (Great Sewerage Canal) used to function by gravity. However, this system was disrupted by subsidence, and, by 1950, the uneven settlement of the sewerage network made it necessary to pump wastewater from the small sewerage lines to the level of the main wastewater collector of the City, thus significantly increasing both maintenance and operation costs. The Great Canal has been affected by land subsidence so much, that at present the first 20 km have almost totally lost their inclination. In addition, continually increasing the population in the metropolitan area has rendered the wastewater collection and treatment capacity insufficient.

Accordingly, in 1967, a decision was taken to build another main collector for wastewaters for both Mexico City and the State of Mexico as a combined sewage and stormwater network (Deep Tunnel Sewerage System (hereafter noted as Deep sewerage)). A system of 60 km of sewerage interceptors and deep collectors were constructed along with a new artificial exit from the basin of Mexico in 1975. By 1997, there were 153 km of tunnels in operation. The Deep Sewerage had to be constructed up to 200 m below the ground level to ensure that it will not be affected by land subsidence (DGCOH, undated, 1990; Domínguez, 2000).

The Deep Sewerage has more than 80 interceptors and carries an average annual flow of 48 m^3/s of wastewater and 14 m^3/s of stormwater through primary and secondary networks. The primary network is 50 km long and 6.5 m in diameter, and it is connected to the secondary network, transporting municipal and industrial wastewater, and stormwater through 3.1 m to 5 m diameter tunnels (INEGI, 1999). The Deep Sewerage system stores, transports and disposes wastewater and stormwater through four artificial channels located at the northern end of the basin of Mexico. The system includes 66 pumping stations, regulatory tanks for flow control, storm tanks, 111 km of open canals, rivers which are now used for transporting wastewater, 16 dams and lagoons. The average volume of wastewater and stormwater that is discharged into the ZMCM sewerage system is 2897 MCM (INEGI, 2001). In 2004, this was 2260.23 MCM, of which less than 10% is treated. No information about the percentage of wastewater that is treated in the State of Mexico is available.

A new interceptor was constructed during the period 1998–2000 for the Great Canal. It is to transport stormwater from Mexico City downtown by gravity and thus alleviate the threat of floods in this part of the City. The interceptor is a 1000 m long and 3.1 m diameter-tunnel built 20 m below the ground level, with a capacity of 35 m^3/s (DGCOH, undated, 2000).

Since the City is located within a naturally closed hydrologic basin, it is especially vulnerable to floods. Throughout history, artificial channels had to be constructed to take wastewater and stormwater from of the City. The rainy season in the metropolitan area is characterized by high intensity storms of short duration. The average annual rainfall in the City is 800 mm: 500 mm in the eastern part and around 1000 mm towards the southern and western parts (Domínguez, 2000). The main collector of the Deep Sewerage was designed to carry about 200 m^3/s of water over a 45-hour period. However, it has carried up to 340 m^3/s. Such sudden fluctuations in the amounts of water that have to be drained create major operational and maintenance problems.

The floods in Mexico City can be explained due to the difference in levels between some parts of the City and the Great Canal, as well as the inability of the sewerage system to quickly pump out all the water during the rainy seasons. For example, due to the subsidence in the City, downtown is 7 m below the highest point of the Great Canal (Legorreta *et al.*, 1997). Since the secondary sewerage network is insufficient to carry high volumes of storm and wastewater, severe problems have been encountered in those parts of the City that are above the east interceptor where the Great Canal has lost its gradient. On many occasions wastewater has also flooded the streets in these areas, but for only for short durations.

Some 30 years ago, the Great Canal could discharge 90 m^3/s. At present, it discharges only 12 m^3/s. Due to this increasing inefficiency, the Deep Sewerage did not receive proper maintenance until 1995, when the heavily silted primary sewerage network could be cleaned. In May 2005, the Water System of Mexico City initiated monitoring activities to check the status of different sections of the main sewerage network of the Deep Sewerage, especially in terms of the infrastructure and level of siltation. The risks presented by sulphuric acid and methane to human beings were nullified by adding a chemical (*Albisol*) to reduce both the acid and the methane to negligible levels. The main findings were that the percolation of water to the tunnels was minor, that the concrete walls of the tunnel had not deteriorated seriously, and that siltation was not serious enough to prevent water from flowing out the network system.

The Master Plans for Drinking Water and for Sewerage for the Mexico City (DGCOH, 1997a, 1997b) outlined the different types of strategies, including infrastructure, necessary to improve the supply, storage and transportation of drinking water in the City, as well as the storage, transportation and disposal of wastewater and stormwater out of it. However, these plans also noted that, in addition to very high investment costs, the infrastructure would also require several years to construct. This means that in spite of the importance of the infrastructure as part of a water and wastewater management strategy for the Mexico City, this is not the only alternative available.

One example is the so-called 'reuse' of wastewater produced in the ZMCM. The disposal of untreated wastewater has become a serious problem for the metropolitan area, especially when the high volume and the nature and levels of pollutants contained therein are considered. The problems created by the current effluent disposal practices are now affecting neighbouring areas of the region, where wastewater is discharged. This has created very significant health and environment-related problems and concerns.

Globally, ZMCM is now by far the largest single producer and exporter of wastewater that is used for agricultural purposes. Since the beginning of 20th century, wastewater from the

City has been diverted to the Mezquital Valley, in the nearby state of Hidalgo, located 109 km north of Mexico City. Otherwise a semi-arid region, the Valley has become an important agricultural area by using this untreated wastewater, with 110 000 ha of official and unofficial command area, and more than 50 000 water users in the different irrigation districts.

In the Mezquital Valley, the main crops grown are alfalfa and maize, representing some 60–80% of the total irrigated area. Cultivation of higher-value crops is forbidden by law due to health considerations. This practice of wastewater irrigation has provided added nutrients to soils and it has been a source of water for economic activities. However, for many years, it has also represented a very high risk to the health of not only the population who live and work in the irrigation districts, but also to the consumers (IDRC, 2002).

In 1996, the Inter-American Development Bank approved a $1.035 billion project for the Mexico Valley Sanitation Project. Unfortunately, this much-needed project did not proceed for several reasons, mainly economic and political. In 2004, the Mexico City Water System, Water Authorities from the State of Mexico and National Water Commission were working jointly with the Inter-American Development Bank and the Japanese Bank for International Cooperation to develop the terms and references to prepare three tenders to build four wastewater treatment plants. The total budget for this project was approximately $1 billion, of which IDB would contribute $365 million for the collectors system and JBIC would provide $670 million for the wastewater treatment plants (STAT-USA, 2004). No public information is available as to what has happened to these projects. The disturbing fact continues to be that more than 60m^3/s of wastewater continues to be discharged with no treatment whatsoever.

The continuous transfer of wastewater over a century and the excessive irrigation by the farmers in the Mezquital Valley to counteract its salinity, have resulted into groundwater recharge of the local aquifer. The groundwater level table has gone up and several springs have appeared, which have become a source of water for the local population. Unfortunately, no serious and reliable study is currently available on the quality of groundwater or the springs in the Valley, as well as their overall impacts on human health and the environment.

Clearly, long-term and rational planning is urgently needed in the ZMCM including an efficient systemic strategy for drinking water and wastewater management. There is an urgent need to formulate coordinated policies for the development and management of the metropolitan area as a whole. To date, there are no signs that this is likely to occur in the near future. As the National Population Council (CONAPO, 2000, p. 79) has noted: "there is no long-term planning for the ZMCM in terms of urban development, including provision of services such as housing and infrastructure". This lack of systemic planning is contributing to increasingly disorganized development of the metropolitan area, which will require a never-ending provision of services such as water supply and sanitation. In addition, technical, managerial and administrative capacities to provide such necessary services is simply not available at present.

Water Pricing Policies

In the ZMCM, drinking water is charged per cubic metre and its price increases with the highest consumption levels. Within the metropolitan area, there is no uniform policy for water pricing. It is decided independently by the governments of Mexico City and the State of Mexico, and even by the few water utilities that operate in some of the municipalities in the State of Mexico.

One of the main problems for the local governments in terms of cost-recovery has been that there were, and still are, numerous water connections that are not registered, and thus consumption through them is neither recorded nor charged. In 2000, it was officially estimated that there were about 2.5 million water connections in the ZMCM: 67% domestic, 16% commercial and 17% industrial (INEGI, 2001). However, these figures represent only approximately 64% of the existing connections, the rest are illegal.

Another reason why water consumed is not charged in the metropolitan area is because most houses do not have meters. In fact, only 49% of the legal connections are metered. In addition, water users currently pay only 24% of the operational, maintenance and administration costs. It is estimated that in 1997, only 43.6% of water was billed at the national level (INEGI, 2000b).

In the case of the Mexico City, the local government has recognized the limitations it faces to provide water to its population. Some of the main problems that have been identified include a deficit of water availability of 3000 l/sec; leakages of more than 30% because of poor conditions of the networks; unreliable water supply received by at least 1 million people; the number of people with no access to water is increasing; and, as of January 2004, Mexico City had not had new sources of water for the previous eight years (Gobierno del Distrito Federal, 2003).

Even though the government expects the implementation of stage four of the Cutzamala System, it has acknowledged that this project, if and when it is implemented, will take several years to be completed. This situation forced the government of Mexico City to develop a strategy to improve the current situation in terms of providing drinking water to the City. The importance of using economic instruments to improve water and wastewater management is slowly being realized. Water can no longer be considered as a public good that is to be supplied by the State to all the users at highly subsidized prices. The strategy for water management currently includes legal and institutional reforms; participation of the private sector for specific activities, such as billing, meter-reading and leakage repairs; and modifications of the pricing mechanism.

A census of water users was carried out between 1994 and 1996. It included all properties and taps that existed in the 16 boroughs. Users were identified and a users' register was prepared. In addition, water consumption was measured (which was virtually non-existent before) by installing meters in more than 90% of the properties. Even though Mexico City has not received additional volumes of water since 1995, the programmes of meter-reading and detection and repair of leaks claim to have saved 2.8 m^3/s of water, with which it has been possible to provide more people with drinking water (Marañón, 2004). Table 4 shows some indicators that illustrate the improvements that have been recently achieved.

In terms of perceptions of the users on the quality of water services, the main complaints are about poor water quality, reliability of the service, and pricing (Marañón, 2004). While the differences in opinions may be considered normal within such a large population, the fact remains that there is a very high percentage of poor people in the City and hence particular attention must be given to the problems to ensure that poor people have access to water in a fair and equitable manner.

The evolution of the tariffs structure for the domestic sector is shown in Table 5.

For the State of Mexico, drinking water is also priced volumetrically. Prices also increase with higher consumption levels, as is the case for Mexico City. However, in some municipalities, the tariffs also vary by area depending upon their dominant socio–economic

Table 4. Efficiency indicators for drinking water supply in Mexico City, 1996 and 2001

Indicators	1996	2001
Volume of water delivered	686.6 million m^3	752.2 million m^3
Volume of water produced	1096.9 million m^3	1087.0 million m^3
Number of meters installed	737.2 thousand	1255.9 thousand
Number of users billed	1477.5 thousand	1769.1 thousand
Amount of water billed	$1.1 million	$ 3.2 million
Amount of water that was paid	$1.7 million	$ 3.8 million

Source: Marañón (2004).

Table 5. Evolution of domestic water tariffs, Mexico City, 1996–2002 (1996 = 100)

Consumption (m^3)	1996	1997	1998	1999	2000	2001	2002
10.1–20.0	100.0	96.0	81.0	82.9	76.1	72.9	73.4
20.1–30.0	100.0	172.8	169.4	162.6	149.4	143.1	144.2
240.1–420.0	100.0	293.0	275.5	292.7	306.6	319.7	322.7
420.1–660.0	100.0	582.1	613.3	653.6	684.5	714.0	720.7
660.1–960.0	100.0	978.7	1049.9	1138.1	1192.0	1243.0	1254.7

Source: Financial Codes of Mexico City, 1996–2002; and National Index of Prices for the Consumers, Bank of Mexico, in: Marañón (2003, 2004).

conditions. Various socio-economic strata have been defined within each municipality, and the people at the higher strata have to pay higher charges compared to the lower strata.

The overall efficiency of water management in the municipalities that are part of the ZMCM is considered to be very low. For example, only 22.5% of domestic consumers pay for water as do 48.7% of non-domestic consumers. Furthermore, water charges in the State of Mexico are still based primarily on fixed rates. Accordingly, demand management practices have been mostly ignored by the institutions concerned. Table 6 shows selected efficiency indicators for 14 municipalities of State of Mexico, where information is available.

Overall, neither Mexico City nor the State of Mexico have carried out serious studies on tariff structures. Accordingly, pricing has played only a minor role for managing water in the ZMCM.

Table 6. Efficiency indicators for 14 water utilities in the municipalities of the State of Mexico which are part of the ZMCM, 2002

Average rate for domestic users (metered)	$0.5/m^3
Amount billed	$44.5million
Average rate for domestic users (fixed rate)	$0.5/m^3
Amount billed	$132.7 million
Non-domestic users (metered service)	$2/m^3
Amount billed	$32.9 million
Non-domestic users (fixed rate)	$4.3/m^3
Amount billed	$134.8 million
Total payment by domestic consumers	22.5%
Total payment by non-domestic consumers	48.7%

Source: Comisión del Agua del Estado de México (2002), in: Marañón (2004).

There are many constraints to improve access to water, quality of water supplied and overall water services. These constraints include issues such as: management continues to be very centralized, hierarchical and bureaucratic; pricing structures have not been properly developed; management and technical expertise available to manage water and wastewater systems are inadequate; users have very little say on how water is managed; and lack of transparency.

A strategy for water management in Mexico City was launched in 1992 as an effort to promote major structural changes. The idea was that water could no longer be considered as a public good (and, as a result, subsidized heavily by the State), but as an economic good. The institutions concerned faced a severe crisis because of deterioration of infrastructure and economic conditions, inefficiency, and a pricing system based primarily on fixed tariffs. It was also necessary to eliminate heavy subsidies because of financial reasons and also to promote water conservation. In addition, due to economic constraints, it was not possible to expand and improve the supply to the poorest neighbourhoods (CADF, 1993, pp. 2–3 in Marañón, 2004). As part of this strategy, private sector companies were invited to participate in different activities such as distribution of drinking water, metering, billing, customer support and maintenance of the secondary networks.

For detailed analysis of the participation of the private sector in managing water in Mexico City, detailed information can be found in CCE and CMIC, 2000; Marañón 2003, 2004; Martínez Omaña *et al.*, 2002; Sistema de Aguas de la Ciudad de México, 2005; Tortajada, 2005, 2006.

Concluding Remarks

Based on the analysis presented in this paper, it is evident that the management of water resources in the ZMCM is very complex and at present inefficient. There appears to be an uneven race between the water and sanitation needs of an increasing population, and the planning, investments, technology and management needs required to construct, operate and maintain all the necessary systems efficiently.

The problems of water quantity and quality in the ZMCM are multidimensional and are directly linked to the societal expectations, regional economic development policies and steady increases in population. The government policies have attempted to promote the development of other urban centres to alleviate poverty and to provide improved standards of living as well as quality of life. However, even though the population growth rate in Mexico City during the later part of the 20th century has declined compared to the rates witnessed in the earlier decades, the growth rates in the adjacent municipalities of the State of Mexico that are part of the metropolitan area are expected to increase even further. Accordingly, the problem is likely to remain complex in the foreseeable future. Unless the current trends and management practices change, the future solutions will require very high investment costs to transport more and more water from increasingly distant and expensive sources. In turn, there will be important economic, social and environmental implications for the exporting regions, higher land subsidence rates due to ever-increasing groundwater withdrawals, a reduction in the quality of the groundwater abstracted, and higher investments to cover operation and maintenance costs, not to mention the decreasing quality of life of the population living in the region.

One constraint stems from the fact that the demand for living spaces from the continually increasing population has contributed to major changes in land-use practices.

Concrete and asphalt now cover areas that are needed for groundwater recharge. The southern area of the City is a good recharge area since the soil is broken basalt. However, this area is now heavily urbanized, and hence is also one of the main sources of groundwater contamination because of the absence of a sewerage network, which cannot be economically constructed due to the presence of volcanic rocks. Housing complexes are thus built only with septic tanks that are mostly not properly constructed and maintained and, therefore contribute to groundwater pollution.

Changes in land use have also contributed to higher volumes of stormwater discharges to the sewerage system, requiring increasing capacities in the system. The risk of aquifer contamination is enhanced because of disposal of untreated industrial wastewaters directly into the sewerage system, inadequate wastewater treatment facilities, leakages from the sewerage networks, and solid waste illegally dumped in landfills, unlined sewerage canals and watercourses.

There has been evidence of low water quality in the aquifer for several years (National Research Council *et al.*, 1995; UNAM, 1997; Mazari-Hiriart *et al.*, 2000, 2001). Total and faecal coliforms, as well as bacteria responsible for gastroenteric diseases and acute diarrhoeas, have been found in groundwater in the southern and western parts of the City. Some studies show that the highest contamination of groundwater is in the centre of the Mexico City (Mazari-Hiriart, 2001). The gastroenteric diseases, resulting from the consumption of polluted water, are the second major reason for child mortality (278 per 100 000) in the country, the third leading cause of death for children in the State of Mexico (450 per 100 000); and the fourth in Mexico City (157 per 100 000).

A major limitation to analysing water problems of the ZMCM, or indeed any part of Mexico, is that of data reliability and accessibility. Official data are often inconsistent, which is a major constraint for decision-making, since decisions have to be taken based on conflicting or no information. The public have very limited access to information available at the institutions which, in addition, is often contradictory from one year to another, from one location to another or even from the same source.

It is obvious that the current approach to the management of the water supply and wastewater in the metropolitan area is neither efficient and equitable, nor sustainable. In order to fulfil the needs of an expanding population in terms of water quantity and quality, and to simultaneously maintain a proper balance between the people, natural resources, environment and health, it is necessary to formulate and implement a long-term integrated management plan, which does not exist at present. This should explicitly consider the needs and interests of the different social and economic sectors in both Mexico City and State of Mexico, and also the numerous existing inefficiencies in management can be overcome. Water allocations for the different consumers need to be systematically planned and be better organized. More efficient institutional arrangements and coordination between the governments of both the regions of ZMCM are essential. Joint and more efficient institutional mechanisms are needed to substantially improve the existing practices. The relevance and importance of public involvement and consultations in preparing and implementing such plans should not be underestimated. Such stakeholders' participation is now conspicuous by its absence.

The current policies on tariff structures need to be reassessed. At present, there is one tariff, based on the volume of water consumed, for the great majority of the people, irrespective of their socio-economic status or the place where they live. Since a poor family can have 10 people living in the same house, they often pay more than a rich family of only 2–4 persons. A new, realistic and equitable tariff structure needs to be developed which will promote water conservation, improve the financial self-sufficiency of the water institutions and explicitly consider access to water by the poor, perhaps with targeted subsidies.

Finally, there is no doubt that there is an enormous potential for improvement in the existing and proposed practices for water management in the Metropolitan Area of Mexico City. However, a policy that considers exclusively the water sector is unlikely to be successful. It needs to concurrently consider linkages to policies on urban development (so far an issue that has been completely ignored), migration, industry, energy and the environment. It will not be an easy task, but nevertheless, it is an essential task.

Acknowledgements

The author acknowledges the support of Dr Germán Martínez-Santoyo, Director General of the Water System of Mexico City (Sistema de Aguas de la Ciudad de México) for the preparation of the paper. Dr Juan Manuel Anguiano-Lozada, Head, Deep Sewerage, Water System of Mexico City, and Dr Nancy Contreras-Moreno, Professor, National Autonomous University of Mexico, provided invaluable support and insightful comments on this paper. The support of Dr Yosu Rodríguez-Aldabe, Advisor to the Director General, CentroGEO 'Ing. Jorge L. Tamayo', National Research Council for Science and Technology, is also acknowledged. CentroGEO prepared several figures specifically for this analysis.

This paper is based on a much longer version that was prepared for 2006 Human Development Report of the United Nations Development Programme.

Notes

1. This section is based mainly on information from National Population Council, Demographic and Urban Scenarios of the Metropolitan Area, Mexico, 2000.
2. The Law of Urban Development of Federal District was published on the Official Magazine of Federal District (Gaceta Oficial del Distrito Federal) on 29 January 1996; and on the Official Newspaper of the Country (Diario Oficial de la Federación) on 7 February 1996. It has been modified three times: on 23 February 1999 and published on Gaceta Oficial del Distrito Federal. No. 25; on 29 January 2004, published on Gaceta Oficial del Distrito Federal No. 8-TER, and on 29 January 2004, published on Gaceta Oficial del Distrito Federal No. 8-TER.
3. Figures include only the municipalities of the State of Mexico where information was available.

References

CAEM (2003) *Prontuario de información hidráulica del Estado de México* (México: Comisión del Agua del Estado de México, Secretaría del Agua, Obra Pública e Infraestructura para el Desarrollo, Gobierno del Estado de México).

CCE and CMIC (2000) *El desafío del agua de la Ciudad de México* (México: Consejo Coordinador Empresarial/Cámara Mexicana de la Industria de la Construcción).

CNA (undated, a) *Subsistema Chilesdo, tercera etapa Sistema Cutzamala. Gerencia de Aguas del Valle de México* (México: Unidad de Información y Participación Ciudadana, Comisión Nacional del Agua).

CNA (undated, b) *Sistema Cutzamala, ramal norte macrocircuito, I Etapa. Gerencia de Aguas del Valle de México* (México: Unidad de Información y Participación Ciudadana, Comisión Nacional del Agua).

CNA (undated, c) *Sistema Cutzamala, ramal norte macrocircuito, II Etapa. Gerencia de Aguas del Valle de México* (México: Unidad de Información y Participación Ciudadana, Comisión Nacional del Agua).

CNA (undated, d) *Sistema Cutzamala, ramal norte macrocircuito, III Etapa. Gerencia de Aguas del Valle de México* (México: Unidad de Información y Participación Ciudadana, Comisión Nacional del Agua).

CNA (1997a) *Diagnóstico ambiental de las etapas I, II y III del Sistema Cutzamala* (México: Comisión Nacional del Agua).

CNA (1997b) *Manifestación de impacto ambiental modalidad específica del proyecto Macrocircuito Cutzamala* (México: Comisión Nacional del Agua).

CNA (2005) *Water Statistics in Mexico* (Mexico: Ministry of Environment and Natural Resources/ National Water Commission).

CONAPO (2000) *Demographic and Urban Scenarios of the Metropolitan Area of Mexico City, 1990–2010* (Mexico: National Council for Population).

DGCOH (undated) El saneamiento del gran canal del desagüe. Dirección General de Construcción y Operación Hidráulica (México: Departamento del Distrito Federal).

DGCOH (1990) El Sistema de drenaje profundo de la Ciudad de México, Dirección General de Construcción y Operación Hidráulica (México: Departamento del Distrito Federal).

DGCOH (1997a) Plan Maestro de Agua Potable del Distrito Federal, 1997–2010, Dirección General de Construcción y Obras Hidráulicas (México: Gobierno del Distrito Federal).

DGCOH (1997b) Plan Maestro de Drenaje Profundo 1997–2010, Dirección General de Construcción y Obras Hidráulicas (México: Gobierno del Distrito Federal).

DGCOH (2000) Alcantarillado. Estrategia para la Ciudad de México. Dirección General de Construcción y Obras Hidráulicas (México: Gobierno del Distrito Federal).

Domínguez, R. (2000) Las Inundaciones en la Ciudad de México, Problemática y Alternativas de Solución, Revista Digital Universitaria, UNAM, 1(2), available at www.revista.unam.mx.

Gobierno del Distrito Federal (2003) Abastecimiento de agua potable. Available at www.obras.df.gob.mx

Gobierno del Distrito Federal, Secretaría de Medio Ambiente del Distrito Federal y Fundación Friedrich Ebert (2004) Hacia la agenda XXI de la Cd. de México (México: Fundación Friedrich Ebert).

Gobierno Federal (2005) Estadísticas. V Informe de Gobierno (México: Presidencia de la República).

IDRC (2002) Estudio complementario del caso Mezquital, estado de Hidalgo, México. Convenio IDRC-OPS, HEP, CEPIS, 2000–2002 (Mexico: International Development Research Centre).

IMTA (1987) Visita al sistema Cutzamala. Boletín No. 2 (México: Instituto Mexicano de Tecnología del Agua).

INEGI (1999) *Estadísticas del medio ambiente de Distrito Federal y Zona Metropolitana* (México: Instituto Nacional de Estadística, Geografía e Informática).

INEGI (2000a) *XII Censo general de población y vivienda* (Mexico: Instituto Nacional de Estadística, Geografía e Informática).

INEGI (2000b) *Cuaderno estadistico de la Zona Metropolitana de la Ciudad de México* (México: Instituto Nacional de Estadística, Geografía e Informática).

INEGI (2001) *Estadísticas del medio ambiente del Distrito Federal y Zona Metropolitana 2000* (México: Instituto Nacional de Estadística, Geografía e Informática).

INEGI (2003) *Cuaderno estadístico de la Zona Metropolitana de la Ciudad de México* (México: Instituto Nacional de Estadística, Geografía e Informática).

Legorreta, J., Contreras, M. C., Flores, M. A. & Jiménez, N. (1997) *Agua y más agua para la Ciudad*, Red Mexicana de EcoTurismo. Available at planeta.com.

Marañón, B. (2004) Tariffs for drinking water in Mexico City, 1992–2002: towards water demand management?, in: C. Tortajada & A. K. Biswas (Eds) *Water Pricing and Public-Private Partnership in the Water Sector*, pp. 61–130 (Mexico: Miguel Angel Porrua).

Marañón, B. (2005) Private-sector participation in the management of potable water in Mexico City, 1992–2002, *International Journal of Water Resources Development*, 21(1), pp. 124–138.

Marañón-Pimentel, B. (2003) Potable water tariffs in Mexico City: towards a policy based on demand management? *International Journal of Water Resources Development*, 19(2), pp. 233–247.

Marañón-Pimentel, B. (2004) Participación del sector privado en la gestión del agua potable en el Distrito Federal, in: C. Tortajada, V. Guerrero & R. Sandoval (Eds) *Hacia una Gestión Integral del Agua en México: Retos y Alternativas*, pp. 289–366 (México: Miguel Angel Porrua).

Martínez-Omaña, M. C. (2002) La gestión privada de un servicio público: el caso del agua en el Distrito Federal, 1988–1995 (México, DF: Instituto Mora, Plaza Valdés Editores).

Mazari-Hiriart, M., de la Torre, L., Mazari-Menser, M. & Escurra, E. (2001) Ciudad de México: dependiente de sus recursos hídricos, *Ciudades 51*, julio–septiembre, Puebla, México, pp. 42–51.

Mazari-Hiriart, M., Bojórquez Tapia, L. A., Noyola Robles, A. & Díaz Mondragón, S. (2000) Recarga, calidad y reuso del agua en la Zona Metropolitana de la Ciudad de México, in: Marxos Mazari (Ed.) (México: Dualidad Población-Agua, Inicio del Tercer Milenio El Colegio Nacional).

National Research Council, Academia de la Investigación Científica, A.C. & Academia Nacional de Ingeniería, A.C. (Eds) (1995) *Mexico City's Water Supply: Improving the Outlook for Sustainability* (Washington DC: National Academy Press).

OECD (2002a) *Territorial Reviews, Mexico* (Paris: Organisation for Economic Co-operation and Development).

OECD (2002b) *Economic Surveys, Mexico* (Paris: Organisation for Economic Co-operation and Development).

OECD (2003) *Social Issues in the Provision and Pricing of Water Services* (Paris: Organisation for Economic Co-operation and Development).

Programa de Naciones Unidas para el Medio Ambiente, Gobierno del Distrito Federal, and CentroGeo, Centro de Investigaciones en Geografía y Geomática, Ing. Jorge L. Tamayo, A.C., (2003) *Una visión del sistema urbano ambiental* (México: GEO Ciudad de México).

Rodwan, J. G. (2004) Bottled water 2004: US and international statistics and developments, *Bottled Water Reporter*, April/May.

SEMARNAT/CNA (2000) El Agua en México: retos y avances (México: Secretaría de Medio Ambiente y Recursos Naturales/Comisión Nacional del Agua).

Sistema de Aguas de la Ciudad de México (2005) *Relación de Concesiones otorgadas en el 2004* (México: Ciudad de México).

STAT-USA (2004) Available at www.stat-usa.gov/

Tortajada, C. (1999) *Environmental Sustainability of Water Management in Mexico* (Mexico City: Third World Centre for Water Management).

Tortajada, C. (2001) Water supply and distribution in the metropolitan area of Mexico City, in: J. I. Uitto & A. K. Biswas (Eds) *Water for Urban Areas* (Tokyo: United Nations University Press).

Tortajada, C. (2004) Urban water management for a megacity: Mexico City Metropolitan Area. Presentation made at the Seminar on Water Management in Megacities, *World Water Week*, Stockholm, 15 August.

Tortajada, C. (2005) Mexico City Metropolitan Area: The largest megacity in an unsustainable path. Presentation made at the *"Water in Megacities" Dialogue Forum*, Munich Re Foundation, Munich, 10 May.

Tortajada, C. (2006) Who has access to water? Case study of Mexico City Metropolitan Area. Analysis prepared for the 2006 Human Development Report, United Nations Development Programme.

UNAM (Eds) (1997) *Environmental issues: the Mexico City Metropolitan Area* (México: Programa Universitario del Medio Ambiente, Departamento del Distrito Federal, Gobierno del Estado de México, Secretaría de Medio Ambiente, Recursos Naturales y Pesca).

Megacities and Water Management

OLLI VARIS, ASIT K. BISWAS, CECILIA TORTAJADA &
JAN LUNDQVIST

Introduction

Cities grow ever faster. Cities grow ever bigger. Imagine a middle-sized town of 200 000 inhabitants. With so many people the world's urban population increases in one day. In one month this growth is almost 6 million and in a year around 70 million. This implies that rural areas must supply escalating amounts of food, energy and many other commodities to towns and cities. Mass flows grow rapidly, distorting ecosystems, as do markets, distorting ages-old social systems.

One of the most striking aspects of urbanization is the growth of very large urban centres. Whereas in 1985 there were 31 megacities, i.e. urban agglomerations with more than 5 million inhabitants, the number had grown to 40 by 2000, and this is expected to grow to 58 by the year 2015. Their population growth is equally dramatic; between 1985 and 2015, from 273 to 617 million. The annual increment to megacities is 9 million, which means 25 000 persons per day. Megacities will double in number and population size between 1985 and 2015. This growth is partly the result of a modification of the administrative boundaries, i.e. some of the annual increment of 9 million consists of people who are living in an area that previously was outside the megacity boundary but which has become incorporated into the megacity.

The rapid growth of the megacities in the developing world has posed major water planning and management changes. In 1994, of the 10 largest cities of the world, only three were in developed countries. By 2015, only two, Tokyo and New York, are expected to stay in this list. However, whereas Tokyo's population is estimated to increase by less than 5% during this period, cities like Jakarta, Karachi, Lagos and Dhaka are expected to grow by 60% to 75%.

Urbanization and growth of megacities are not new phenomena: cities such as London or New York started to grow in the 19th century. However, two important differences should be noted which have made the urbanization process (and provision of water supply, sanitation and stormwater disposal services in the megacities) of the developing world fundamentally different from their counterparts in the developed world one century earlier (Biswas *et al.*, 2004).

The first difference is the rate of growth. The development of the megacities in the developed world was a gradual process. For example, much of the population growth in cities such as London and New York was spread over a century. This enabled these cities to progressively and effectively develop the necessary infrastructures and management capacities for all their water-related activities and services.

In contrast, the megacities of the developing world witnessed explosive growth during the post-1950 period, and especially after 1960. For example, the population of Mexico City Metropolitan Area increased from 3.1 million in 1950 to 13.4 million in 1980, a 425% increase in only 30 years. This expansion continues still as the City's population has now exceeded 18 million. Such megacities were simply unable to manage such explosive growth rates. They had to run faster and faster to stay in the same place. The fastest growing megacities are expected to grow more than fourfold in 25 years. Such cities include Dhaka (Bangladesh), Lagos (Nigeria), Guatemala (Guatemala) and Jinxi (China). The dimension of the urbanization development is striking, whether considered either from the standpoint of megacity growth, augmentation of the urbanization level or from the growth rates of urban population.

The second major difference is that as the megacities of the industrialized countries expanded, their economies were growing concomitantly. Accordingly, these urban centres were economically able to harness financial and human resources to provide their residents with the necessary water-related services (Biswas *et al.*, 2004).

In stark contrast, economies of the developing world have mostly performed poorly during the period of this rapid urbanization. High public debts, inefficient resource allocation, poor governance, lack of investment capital and inadequate management capacities have ensured that the necessary infrastructures could not be built on time, and the existing facilities could not be properly maintained.

The urban poverty problem is massive and increasing. Whereas the number of the urban poor is approximated to be around 1 billion today, it is expected to double by 2030. After UN HABITAT (2004), the humble UN target defined in the Millennium Development Goals to reduce the number of slum dwellers to half is likely to remain a dream. What seems more likely is that the number of slum dwellers will double. Slum upgrading policies, particularly in bigger urban agglomerations are therefore expected to become very important in the coming decades, stirring also the water sector. Figure 1 contains UN HABITAT estimations of the percentages of urban slum populations in the countries and regions of the case study cities.

How these excessively large and rapidly growing urban centres manage water in all its aspects is the question that is elaborated and summarized in this paper. The particular focus is how the water sector addresses the present and future challenges and how sustainable the approaches are. This analysis summarizes the situation in eight megacities

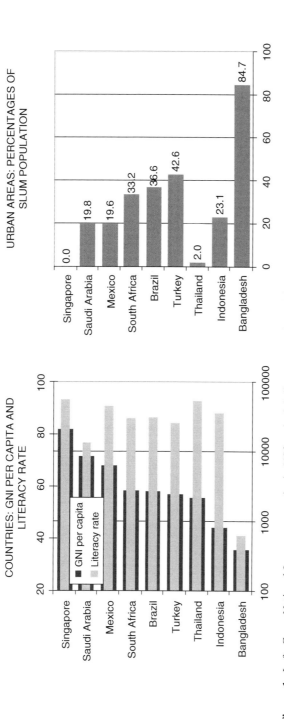

Figure 1. Left: Gross National Income per capita in US$ and adult literacy rate (over 15 years) in 2000 (World Bank, 2004). Right: Percentage of urban slum population by country. *Source:* UN HABITAT (2003a).

worldwide. Those cities are Dhaka, São Paulo, Mexico City Metropolitan Area, Jakarta, Johannesburg, Istanbul, Riyadh and Singapore.

There have been a number of recent analyses on this topic by organizations such as UN HABITAT (2003a), 2003b, 2004) and development banks. However, the authors of this paper feel that the magnitude and importance of this problem is not properly comprehended and the urbanization problem has been absolutely inadequately addressed in recent key events related to water policies, such as the World Summit on Sustainable Development of Johannesburg in 2002 as well as the three World Water Forums (from 1997 to 2003).

Megacities and Water

As the cities grow, their water and wastewater disposal requirements grow as well. Water management was not a serious problem as long as population numbers were low and concentrations of the people were not high. As the population increased dramatically in the last 50 years, and the rate of urbanization began to accelerate, the provision of clean water and safe disposal of wastewater and stormwater in the megacities of developing countries became increasingly more complex and serious (Biswas *et al.*, 2004).

Continuing urbanization poses a major challenge in providing adequate water services to the megacities, but its importance and contribution towards the development of stronger and more stable national economies should not be underestimated. In 2000 it was estimated that the urban areas of the developing world, which contained some 30% of the total population, contributed nearly 60% of the total GDP and played an equally important role in terms of social development and cultural enhancement. Thus, the urbanization process presents both opportunities and challenges.

The main problem of megacities often stems from the fact that the rates of urbanization have often far exceeded the capacities of the national and local governments to plan and manage the demographic transition efficiently, equitably and sustainably. Living conditions are particularly harsh for the large part of the urban population, maybe about a third, who live in areas which are not planned and where public services are lacking or are rudimentary. Quite often the authorities have been reluctant to recognize these mushrooming parts of the cities and have also been hestitant, or unable, to extend social services, including water and sanitation, to the population in these areas. People in these informal areas, however, constitute a pool of labour and they play a very important role in the economic growth of cities. At the same time, the deplorable living conditions are a threat not only to the quality of life in these areas, but to a stable functioning of the city as a whole (Lundqvist *et al.*, 2003). The high growth rates of megacities have simply overwhelmed the limited capacities and resources of the responsible governments at all levels. There is thus an urgent need for additional water and sanitation services, either from government but more probably in partnership with other responsible actors (Biswas *et al.*, 2004).

These developments have resulted in extensive air, water, land and noise pollution, which have major impacts on the health and welfare of the megacity dwellers. The problem is further compounded by skewed income distribution, high unemployment and underemployment, pervasive corruption and increasing crime rates.

Provision of clean drinking water, wastewater collection and disposal and stormwater disposal have now become serious problems for most megacities, ranging from Manila to Mexico City, and Kolkata to Cairo. Indeed, there is also a mounting need to improve services to industry and service sectors. Fortunately, in many urban centres, progress is being made, new and innovative approaches are being successfully applied, and water institutions in certain countries are undergoing radical transformation. Many of these success stories, even with the current information and communication revolution, are mostly unknown and undocumented.

Besides water supply, sanitation, stormwater and waste management, water is fundamental to megacities in several other aspects too. The number of humans exposed to floods tripled from 1970s to 1990s, being around 2 billion today. The major factor behind this development is the congestion of hundreds of millions of people in mushrooming cities on deltas and floodplains of flood-prone tropical rivers. Interestingly, a considerable number of desert and semi-desert megacities are also currently growing at an incredible pace. These cities have the opposite problems with water—they feel scarcity very concretely in their everyday lives (Biswas *et al.*, 2004).

Cities eat enormous amounts of food that they import from countryside, often far away. Megacities alone import as much virtual water as what crosses national borders in all the international food trade. Likewise, megacities import massive quantities of energy. On average, a megacity dweller consumes 5 to 10 more times the amount of energy than the national average. All agendas say that the growing energy production should be covered by renewable sources. In urban conditions, this means practically hydropower or bioenergy. Both of these energy sources rely absolutely on water (Varis, 2006, this volume).

Water in Eight Megacities

The situation and future prospects of the eight rapidly expanding case-study megacities in various parts of the developing world have been summarized in this volume. Those cities are the following (author's name in brackets): Jakarta (Lanti), Dhaka (Haq), Johannesburg (Turton *et al.*), São Paulo (Braga *et al.*), Mexico City Metropolitan Area (Tortajada), Riyadh (Abderrahman, 2006), Istanbul (Altinbilek) and Singapore (Tortajada).

Jakarta

The capital of Indonesia, Jakarta, is the biggest megacity in Southeast Asia. It is located on the highly crowded island of Java, which has around 120 million inhabitants on only 132 000 km^2, making Java one of the most densely populated areas in the world with its 910 inhabitants per km^2.

Jakarta has an official population of 10 million inhabitants, but its actual population is probably higher. The number of seasonal residents and commuters was already 1.15 million in 1985 (UN, 1989). The city quadrupled its population between 1975 and 1995.

Twenty to twenty-five per cent of the housing in Jakarta is on a temporary basis, in so-called kampung areas, and other 4–5% lives scattered on river banks and other comparable plots of land (McCartney, 2004). The kampungs now have a semi-legitimate status.

The water distribution network of Jakarta was originally built by the Dutch to supply water for 0.5 million inhabitants. Although extended, it cannot cover the whole city area. Due to rapid population growth, development in new industries and large-scale real estate construction, there is a rapid increase in water demand.

The water supply services are operated by two privatized companies, the French Ondeo for the Western Sector and the British Thames Water for the Eastern Sector. This

Table 1. Comparison between expected and achieved indicators of water supply in Jakarta after privatization in 1998

	Condition in 1996	Condition expected in 2002	Achievement in 2002
Groundwater over-abstraction	Severe	Reduced	No significant reduction
Service coverage	41%	70%	44.2% (West) 62.2% (East)
Share of non-revenue water	57%	35%	44%
Water sold	176 million m³/year	342 million m³/year	255 million m³/year

Source: Lanti (2006, this volume).

arrangement has been operational since 1998 when the Provincial Government of Jakarta decided to privatize its water supply operations.

In 1996 the plan was that the private operators would increase the water supply coverage from 40% to 70% of the households by 2002. However, this target remained very far from reality, as did the other targets (Table 1).

Several reasons contributed to this failure to meet goals, including the following:

- The Asian financial crisis, which cut the value of the local currency remarkably against foreign currencies.
- Restrictions to adjust the tariffs for full cost recovery. Despite this, the tariffs went up by 2.5-fold in 2001–04.
- Ambiguous status of former and present government workers in the private companies.
- Lack of fulfilling the obligations on both the government and the private side.

During this period, the average per capita consumption remained at 156 litres per day.

The domestic waste disposal system is highly underdeveloped in Jakarta. The coverage of the sewerage service was to a mere 1.9% of the population in 2001. The only wastewater treatment technology used in this context consists of aerated lagoons. The city has about 1 million septic tanks that cover around 39% of the population, 20% use pit latrines and the remaining 59% discharge their waste directly into dikes, canals and rivers. Highly polluted canals and rivers are used widely as a water source for cooking and washing (Lanti, 2006, this volume).

The extension of the water supply faces severe problems. Over-abstraction of groundwater has already caused remarkable saltwater intrusion into the main aquifers. Surface waters, which are utilized in the water supply, are often so polluted that they are considered worse than the water running through sewage treatment systems in many countries. Network leakages are huge. The water supply companies must make profits, so they focus extensions of the network to households who can pay. Standpipes on kampungs have received attention by the governments but progress is slow. Where there are standpipes, there are often illegitimate vendors selling water to the poorest urban households at over 10 times the official price (Lanti, 2006, this volume).

The area is prone to seasonal floods that raise water into the streets. Wide-reaching groundwater pollution has been observed due to poor waste management. As a response, existing drains have been re-directed in some locations to provide a faster passage of the water into the sea. Pilot scale studies for the construction of a sewer system have already been made since the 1980s (UN, 1989) but not much has happened.

Dhaka

It is not easy to imagine a more water-affluent megacity than Dhaka. Dhaka receives 2000 mm of rainfall annually. It is located close to the convolution of the mighty Ganges, Brahmaputra and Meghna rivers and it is frequently flooded, often catastrophically. These three rivers constitute the world's second biggest river system with an annual discharge 25 times that of the Nile. However, Dhaka is one of the most challenging megacities in its water management.

Dhaka is the political and economic centre of Bangladesh. The country has more than 130 million people in an area of 147 540 km^2 making it extremely crowded. Dhaka's population is approaching 15 million with a growth rate of around 5% per year. Bangladesh is one of the poorest countries of the world with 44% of people living below the poverty line. It has been estimated that around one-quarter of Dhaka's population live in slums (UN HABITAT, 2003b).

The water supply and sewerage services have been allocated to one single public authority. It now supplies 0.51 km^3 of water per year against the demand of 0.73 km^3, serving around 72% of the city dwellers. The quality of the supplied water is very much in question. Almost 1000 private wells abstract another 0.35 km^3 per year of groundwater, mainly for industrial purposes. Groundwater is used far beyond the sustainable rate and this groundwater mining puts a serious strain on the environment. The groundwater table has gone down 20 to 30 m in the past three decades and continues to sink 1 to 2 m per year (Zahid *et al.*, 2004). Fortunately, the groundwater used by Dhaka is free from arsenic.

Seventy per cent of the population has adequate sanitation and 30% are served by sewer networks. Only one sewage treatment plant exists, with a treatment capacity of 49 000 domestic connections. This is not a great deal for a megacity of the size of Dhaka. Over one-quarter of population lacks adequate sanitation altogether (Haq, 2006, this volume).

The share of unaccounted-for-water is around 53%. It has gradually decreased from the level of 75% in 1980. There are currently important discussions on various water management issues such as the cost recovery of water services through tariff regulation, increased involvement of the private sector to water management etc. in order to bring more efficiency and transparency to the water sector of Dhaka (Haq, 2006, this volume).

Serious surface and groundwater pollution with detrimental effects on public health follow from the massive infrastructure shortcomings in water supply and sanitation. They are reinforced by occasional and often dramatic flooding, which raises the water level to streets and dwellings. Stormwater management systems have been developed but not at pace with the growing population, particularly for the Eastern part of Dhaka with a population of 3 million (Ahmad & Kamal, 2004). Several decades ago, the city was covered by a canal network of 24 canals and included a large area of natural wetland. This system was able to keep flood damage fairly low. The unplanned and largely illegal sprawl of the city ever since has led to the situation in which no proper stormwater infrastructure exists. The most important flood protection system today is the Dhaka Western Embankment which is able to keep about half of the city area virtually flood-free (Haq, 2006, this volume).

Johannesburg

Johannesburg is not a megacity according to the UN definition and according to the UN (2002) statistics with its population of 2.4 million in 2005. However, as Turton *et al.* (2006, this volume) point out, the infrastructure and the water management system that serves Johannesburg is also serving several other cities and settlements that have a total population of around 10 million. The system also serves massive industries such as coal and steel and that part of South Africa accounts for 10% of all economic activity of the African continent. Accordingly, Johannesburg is a part of an agglomeration that can easily be called a megacity.

In hydrological terms, Southern Africa is a very challenging region for sustaining such a large agglomeration of human population and industries, particularly in the location of Johannesburg. Even though the total annual rainfall in the uplands where Johannesburg is located totals reaches up to 600 mm on average, the evaporation losses are 1600 mm and the conversion of rainfall to river flow is exceptionally low (see also Basson *et al.*, 1997). In addition, the rainfall variability is very high which makes exploitation of water still more complicated.

Johannesburg is located on a ridge, which is at the headwaters of two major international rivers, the Orange and the Limpopo. The city is thus not by a major river that it could use, but instead its activities have international hydropolitical consequences. The reason why the city has such a very unfavourable geographical location is due to the fact that Johannesburg originally grew around gold mines (Turton *et al.*, 2006, this volume).

How then is the greater Johannesburg area able to manage its water? The solution is extensive indeed. Several dams have been constructed and rivers have been interlinked. In fact, all major rivers in South Africa are now interlinked and considerable water transfers take place between river basins. This causes growing tensions, even internationally. The most important of these transfer systems to Greater Johannesburg is the one that brings water from five large dams in the Lesotho Highlands, through a channel to the Vaal River and then to the Vaal Dam which is the major source of water of Johannesburg. The distance from Lesotho to the city approaches 500 km. In addition, three other large dams have been constructed to facilitate a sufficient water supply to the greater Johannesburg area.

Rand Water, the company responsible for this system, now supplies around 3.3 km^3 of water per year. The supply capacity has recently grown much faster than the demand and consequently there is a considerable overcapacity of up to 60% of demand in the region. The demand projections have been overly high due to lower economic growth than anticipated, an increase in HIV/AIDS and the success in demand management. Despite improved demand management, the share of unaccounted-for-water is still over 50%. On the other hand, such an overcapacity allows good possibilities for risk management, which is very important in the climatic conditions of Johannesburg area. The area is subjected to several other considerable risks such as health risks from HIV/AIDS, potential conflicts, environmental risks and so forth (Turton *et al.* 2006, this volume).

The high share of unaccounted-for-water is not merely a technical issue. Instead, it reflects the enormous social contrasts of the city, which have their roots in ethnic disparities that are dramatic and existed even before the apartheid era. As one of the many reflections, the crime rate in Johannesburg is higher than in any African city that has been recorded in UNCHS statistics (UN HABITAT, 2004).

The water infrastructure is fairly developed in Johannesburg. In 1995, 80% of dwellings had a piped water as well as sewer connection (UN HABITAT, 2003b).

What about the environment? Environmental problems, particularly groundwater quality problems due to mines and mine dumps, and surface water problems due to municipal wastewaters, are all escalating. In the South African part of the Limpopo Basin, for example, there are over 1000 abandoned mines that are a source of heavy metals, sulphur and associated acidity problems etc., which yield serious local and international problems to water users and to the environment. Municipal wastewaters are the cause of growing eutrophication, which have equally wide-reaching effects downstream.

All in all, the Johannesburg case challenges the notion of river basin management very clearly. A great deal of technical innovation has been used to provide this endemically water-scarce city with sufficient water over large distances. However, the social and environmental problems are not solved and in fact they are escalating. At the same time, the dependency and, in a way, the vulnerability of the Johannesburg area to problems with this highly technical system, has grown extremely high. Turton *et al.* (2006, this volume) point out that the future is governed by very high uncertainties and the ingenuity of people to be able to adapt to conditions, which may change enormously in a very short time, is the key of the successful survival of this urban agglomeration.

São Paulo

In one century, São Paulo has grown sixtyfold, from a city of 300 000 inhabitants to a megalopolis of 18 million. It has become the major economic centre of Brazil, accounting for 27% of the national GDP. Like Johannesburg, this megacity is also located in highlands, but it enjoys 2.4 times the rainfall of Johannesburg. Therefore, it does not have a water scarcity problem of the same dimension as Johannesburg. However, due to the enormous size of this megacity, and to its location in a relatively small basin of the upper Tietê river, São Paulo also is totally dependent on large-scale interbasin water transfers. The first of these systems is now more than a century old. Since then the water and energy requirements of São Paulo have incubated the interlinking of several rivers as well as the construction of a high number of reservoirs. This construction work keeps going at a high pace. Crucial to the energy supply of São Paulo are the hydropower plants that make use of the 750 m high slope from the highlands to the coast in the vicinity of the metropolitan area (Braga *et al.*, 2006, this volume).

As in Johannesburg, one single utility, SABESP, operates this huge infrastructure. At present, it provides 67 m^3/s of drinking water to the city, which is 2.1 km^3 annually. The system includes 8 water treatment plants, 1472 km of main canals, 159 pumping stations, 198 urban reservoirs and 25 000 km of distribution lines (Braga *et al.*, 2006, this volume).

The surface area of the upper Tietê river basin is concurrently about 44% urban. The hydrology of the basin has been totally modified from its natural conditions and the consequent environmental problems have become massive. Floods are a serious problem and they constantly worsen as the land becomes less permeable, while at the same time people keep developing settlements on the floodplains. The design flows have grown dramatically, for instance in the Tamanduatei river they have grown fivefold in less than a century. Heavy building work continues to construct more retention basins and, combined with the basin restoration plans, the flood problem is being addressed in a substantial way.

However, the flood problems continue to be extremely critical throughout the wet season from December to March, and the people are therefore frightened.

The first wastewater collection networks and treatment plants were constructed to São Paulo in the 1950s. However, they are now totally insufficient because they have not been extended to keep pace with the city growth. The environmental conditions of the Tietê river became unbearable in the 1980s. Environmental movements and the media, including several AM radio stations, became powerful enough to influence the policies in two ways. First, some of the environmental problems of the São Paulo area were transferred downstream and a public voice in those areas grew accordingly. Second, the so-called Tietê project began, with the aim of constructing sufficient wastewater treatment capacity to cut substantially the effluent load to the receiving waters. The project has now resulted in a situation in which around one-third of the wastewaters are treated and the plan is to have sufficient capacity to treat the wastewaters from 18 million people by 2015. However, by that time the city will be far bigger than 18 million.

Braga *et al.* (2006, this volume) point out the importance of having an integrated approach to water management in megacities such as São Paulo. The traditional, fragmented approach has proved to be insufficient. Even though there is a body of knowledge, skill, an institutional framework and further capacity to deal with water and sanitation, another to deal with flood management, and still another with housing or with transportation, etc., there must be integration and interaction between all of them.

In this respect the recent developments in São Paulo shows positive signs. The politically influential class is more receptive to open discussion and to the advice of professionals and the public than before. The educational level has increased substantially. However, in order to successfully meet the excessive environmental and social problems due to the ever-growing megacity of São Paulo, the institutional challenges are tremendous (Braga *et al.*, 2006, this volume).

Mexico City Metropolitan Area

The Mexico City Metropolitan Area, which includes Mexico City and more than 30 municipalities of the State of Mexico, is 2240 m above sea level. According to the UN (2002) it had 18.1 million inhabitants in 2000, but a more realistic figure is 24 million (Tortajada, 2003). The region receives relatively little rainfall, only just over 700 mm per year. Hence, the provision of an adequate water supply calls for a massive infrastructure including water transfer systems and exploitation of groundwater, both on a huge scale. The rapid growth of this area has caused a decline in the quality of life, becoming overcrowded, immensely polluted and has a serious shortage of basic amenities such as water.

The Metropolitan Area falls under two federal administrative areas, the Federal District (DF), and the State of Mexico. In the former, 93.5% and in the latter, 84.2% of population are served with water, either with direct house connections or common faucets in the neighbourhood (INEGI, 2000). Daily water supply per capita in these areas is 230 and 297 litres respectively. The actual supply is significantly less because this figure includes the water used by industries and services as well as leakages of over 40% and various unauthorized uses. The eastern part is particularly short of a water supply.

More than 5% of people still have to buy their water from water trucks. These poorest people use 6 to 25% of their income to purchase their daily water. The price they have to pay for water is around five times higher than those who have a registered domestic connection.

The water used in the city depends predominantly on local groundwater aquifers and the supply from long-distance surface water transfer systems. Each year $0.9\,km^3$ is abstracted from local sources including groundwater, springs and snowmelt water. The other important sources of water are the Lerma-Balsas and Cutzamala river basins. The water supply system has 16 dams with a total storage capacity of $0.2\,km^3$.

Mexico City Metropolitan Area produces an annual average of 2.3 million m^3 of wastewater. Although 65 treatment plants have already been installed, a mere 9% of wastewater is being treated.

The National Water Commission, together with the governments of the Federal District and the State of Mexico, are responsible for organizing the huge water infrastructure of the Mexico City Metropolitan Area.

Tortajada (2003) points out that despite of the immense infrastructure, no strategies for the integrated management of water resources exist. As a result, severe shortcomings in maintenance and systematic infrastructure development subsist. Occasional heavy rains have caused some rivers—now open sewers—to flood in inhabited areas and it is usually poor people who are exposed to health risks. Equally, cracks in sewer canals have caused wastewater to flood in dwelling areas.

A long-term regional strategy is urgently needed. The involvement of all stakeholders in the development of a strategy for sustainable development of the Metropolitan Area is seen as a necessity by Tortajada (2006b, this volume). Such a plan should include socio–economic development, poverty alleviation, quality of life and water and wastewater management.

The federal government has promoted the growth of other urban centres in the country, but with limited success. The baseline is that the construction of an infrastructure with an appropriate plan to bring more and more water to the Metropolitan Area is not a socio–economically feasible and environmentally sustainable solution. The costs are rocketing and the benefits will predominantly go to the well-off proportion of the population. The neglected environmental deterioration in terms of lowering the groundwater level, land subsidence and degradation, as well as deterioration of surface and groundwater quality, works against the sustainability of this megacity.

Riyadh

According to the United Nations (2002) projections, the capital of Saudi Arabia, Riyadh (or ArRiyadh) is one of the most rapidly growing cities of the world, and it is fast approaching the milestone of a population of 5 million (Varis, 2006, this volume, Figure 5). Since 1970, its population has grown tenfold. Such a growth rate introduces remarkable challenges to all aspects of city management, which in Riyadh's case are multiplied by the harsh climatic conditions since it receives only around 100 mm of precipitation each year. It has no river, nor coast. UN HABITAT (2003a) estimates that 20% of the Saudi Arabian urban population lives in slums.

Riyadh has been one of the hubs of Saudi Arabia's rapid economic development and growth in the past three decades. The massive oil revenues have lubricated this growth; the country is the world's biggest exporter of crude oil with over 15% of global exports (IEA, 2003) and the fossil fuel trade brings around US$60 billion annually to Saudi Arabia as export earnings (UNDP, 2004). These earnings allow very different solutions for the water sector, in comparison to most of the cases in this volume.

It is very common that economic growth results in decreasing population growth. In Saudi Arabia, this has not been the case and the Kingdom's population has grown threefold since 1970. The growth is expected to continue, and doubling of the population is expected within the next 20 years. With an increase of the share of the urban population, this inevitably means a remarkably rapid expansion of urban centres. The urban population of Saudi Arabia was 3.7 million in 1970, 15 million in 2004, and this is projected to escalate up to 33 million by 2025 (Abderrahman, 2006, this volume). Domestic water demands have grown almost tenfold from 1970, and they keep mounting.

Riyadh's local Water Directorate has successfully managed to cope with the huge challenges that come with domestic water supply since it covers around 98% of the population of the city. Riyadh consumed $0.56 \, km^3$ of water in 2000. For decades the local groundwater resources have been insufficient to meet the demands of the city. Today, around 47% of raw water is extracted from groundwater, but most of the water is pumped considerable distances from distant groundwater fields, for example, the newly developed Al Honai fields are at a distance of 218 km from the city. A considerable amount of the groundwater is too salty to be used without a reverse osmosis treatment, and the aquifers exploited are predominantly non-renewable. Riyadh gets over half of its water from seawater through a large-scale desalination plant at Al-Jubail on the Arab Gulf coast. The desalinated water reaches the city through pipes of 466 km of length.

These solutions are highly expensive, both in terms of investment and operation. The Water Directorate receives only 2.5% of its revenues from water tariffs, and has considerable challenges in financing the extension of water supplies in order to meet the growing demand. By 2022, it has been estimated that the finances needed to meet the rising demands of Riyadh will reach US$29 billion. Clearly, the water system in Riyadh is only sustainable as long as the enormous necessary finances are available.

Whereas the city has been successful in providing water supply services for almost all of its dwellers, the situation is not the same for sanitation and wastewater treatment. The sewerage network covers one-third of the city's territory for 56% of the population. Three tertiary-level wastewater treatment plants are in use and they have a capacity of $0.147 \, km^3$ per year, which accounts for one-quarter of the supplied water.

As well as struggling with the enormous financial and technical challenges to supply water to the expanding desert city and treating the used water, the contemporary water policies include notable measures to enhance the recycling of used water and using an increasing share of it in agricultural production. The policies will streamline and centralize the governance system to become more functional, through water awareness and capacity building campaigns and programmes to the public and the professionals, by reducing leakages and managing the demands for water. They will also address the adverse effects of the water use for the environment, particularly to the shallow aquifers in the vicinity of the city (Abderrahman, 2006, this volume).

Istanbul

The historical city of Byzantium-Constantinople-Istanbul has a history spanning 2700 years and the history of the innovative water supply systems are equally as old.

The population of the city passed the 1 million milestone 100 years ago, and after that Istanbul's population doubled within 70 years (Standl, 2003). However, it only took a further 13 years for the population to increase by another 1 million. Since 1980 Istanbul's population growth has been phenomenal. The population doubled from 2.8 to 5.5 million in only five years, and has thereafter needed only four years to attract 1 million more inhabitants. Most newcomers are migrants from Anatolia. Istanbul now has 11.5 million people, and is projected to approach 15.5 million by 2040 (Altinbilek, 2006, this volume). The city is the economic centre of Turkey; 40% of Turkish industry is located in Istanbul.

The expansion of the city in the past three decades has strained the water infrastructure in a notable way. As the city grew an astronomic 14.6% per year in the early 1980s, which was far faster than expected, and growth slowed down at only a modest pace, Istanbul faced severe water shortages in the early 1990s. Turkey still has a momentous slum problem due to rapid expansion of the cities; around 43% of the country's urban population lives in slums (UN HABITAT, 2003a).

In terms of water supply, Istanbul has a challenging location both geologically and geographically. It has an unfavourable geology for aquifers and a hilly terrain. The city lies on both sides of the Bosporus Strait, partly in Asia (around one-third of the population), and partly in Europe. The water supply of the city was already based on the use of large aqueducts in Roman times; the largest system, which was finished in the year 324, was 242 km long. The Roman systems have been in ruins for several centuries. However, many of the Ottoman systems that were constructed from the 15th century onwards are still operational and the largest of them has a capacity of 17,334 m^3/day. This water feeds public fountains, wells, water tanks and public Turkish baths (Altinbilek, 2006, this volume).

The water infrastructure was gradually expanded over centuries. Through the construction of dams and pipelines, the city received 0.41 km^3 more water in 1974 than it did in 1884. However, after that the population expansion led to severe water shortages. Between 1974 and 1994, the capacity only grew by 0.16 km^2, but the population grew by 5.8 million. This means an increment of 76 litres per newcomer per day, which is by far too little for urban uses. The city planners of Istanbul now design the systems for a consumption level for 250 litres per day.

However, Istanbul's water infrastructure saw a massive investment and improvement after that. The supply capacity has been doubled by the considerable extension of the large-scale water transfer systems, particularly on the Asian side. This improvement can at least partly be attributed to the Istanbul Water and Sewage Administration, which was established in 1994, and as well as rapidly implementing new transmission lines, has since been able to improve the quantity and quality of water supplies significantly, and modernize corroded systems which has reduced leakages (Standl, 2003). The share of unaccounted-for water has decreased from over 50% to 34%, which in Istanbul's case includes many public water uses by mosques, public baths, cemeteries and fire hydrants, besides losses due to leakages etc. With this definition, the target for the city's unaccounted-for water is not zero or close to zero but around 25% (Altinbilek, 2006, this volume).

The sanitation and wastewater treatment infrastructure has also undergone a rapid extension and improvement in the past 10 years. In 1993, only 9.3% of the city's wastewaters were treated. By 1996, the number had gone up to 16%, and through a very extensive investment programme, a level of 95% had already been achieved in 2002.

The volume treated grew almost tenfold between 1993 and 2004 (Altinbilek, 2006, this volume).

Istanbul provides a very encouraging success story of how the vast capacity problems of water and sanitation services of a rapidly expanding megacity can be turned into a highly functional system with a high level of services, technology, environmental gains and public acceptance (see Eroglu *et al.*, 2001). It has required institutional reform and the setting-up of an independent public agency (Istanbul Water and Sewerage Administration, ISKI), a great deal of political commitment, massive funding (facilitated by a new tariff structure and additional public funding), as well as technological and other implementation capacity.

Singapore

Another success story is that of Singapore. This city-state in Southeast Asia has evolved from a low-income harbour town to a showcase of high-level, ultramodern technologies in only a few decades. The water sector has been one part of this development, since, as Tortajada (2006a, this volume) points out, the water management institutions can only be as efficient as its management in other economic sectors.

Singapore now has around 4.3 million people on its islands of $700 \, \mathrm{km^2}$ of surface area. Water is extremely scarce in Singapore because the total renewable freshwater resources of the country are only $0.60 \, \mathrm{km^2}$, which means only $140 \, \mathrm{m^3}$ per capita per year (Segal, 2004). So, Singapore imports around half of its daily-consumed water.

After gaining independence from Malaysia in 1965, the two countries had a dispute over water, but they reached a long-term agreement that is in force until 2011. Singapore would like to extend the deal even beyond 2061 but Malaysia has demanded a much higher price for water than at present (15 to 20 times the current tariff), and so far the question is open.

Singapore's policy has therefore been to increase the domestic water security by various means (Tortajada, 2006a, this volume):

- Supply management
  - Protection of water sources. Catchment management receives a growing emphasis and strictly protected areas today cover almost 5% of the whole territory of the city. Protected and partly-protected areas cover a total of half of the land area, but is expected to increase to two-thirds by 2009.
  - Reuse of wastewater. All of the population are connected to the sewage network and all wastewater is collected and treated. An increasing part of the treated wastewater is reclaimed and further purified by dual-membrane and ultraviolet treatment, and supplied thereafter to industrial and commercial customers. By 2010, this NEWater approach is planned to account for 15–20% of Singapore's needs, since the water can be produced at the cost, which is only 40% of the cost of desalinated water.
  - Desalination: The first desalination plant was opened in 2005 and it has the capacity to produce $0.04 \, \mathrm{km^3}$ per year. The cost per $\mathrm{m^3}$ has been as low as US$0.48. Desalination is expected to increase in the future.
  - Reducing unaccounted-for water. Whereas in most Asian urban centres this quota ranges between 40% and 60%, Singapore has been able to decrease this from 9.5% in 1990 to 5% today.
- Demand management. The refinement of the tariff system has been the main reason for the gradual decrease of unit water consumption.

- Governance factors
  - Human resources. Singapore has developed a professional staff policy, which is seemingly free of the typical problems of Asian utilities, which include political connections, nepotism, narrow and undisciplined expertise and overstaffing.
  - Corruption. Singapore has been able to create a strong anti-corruption legislation and culture.
  - Autonomy. The Public Utilities Board of Singapore manages the whole water cycle of Singapore, and it has developed a holistic policy for the water sector. It has a high level of institutional autonomy with clear-cut responsibilities.

Conclusions

Are megacities sustainable in terms of their highly diverse water related aspects (cf. Varis & Somlyódy, 1997)? At the United Nations World Summit on Sustainable Development in Johannesburg in 2002, it was understood that sustainable development should encompass balanced economic, environmental and social development under the prevalence of good governance and public participation. The case studies open a very broad and diverse spectrum of problems, challenges and solutions (Figure 2).

Sustainability was the baseline of Tortajada's (2006b, this volume) analysis on water management in the Mexico City Metropolitan Area. She concluded that the region is on an unsustainable path, due to the past overemphasis on supply management and on the physical infrastructure. Social, environmental and economic policies have been and remain inferior to the needs, as well as long-term strategies. Integration is dramatically missing in the case of the two Latino giants, Mexico City and São Paulo (Braga *et al.*, 2006, this volume; Tortajada, 2006b, this volume). Riyadh's water sector has evolved in a direction that is extremely dependent on massive financial input, which largely comes from oil export earnings. The vulnerability to financial and trade disturbances is hence elevated. Singapore struggles with the dependency of its hi-tech water infrastructure on imported water, and tries to find sustainability and decrease the political vulnerability of its system through producing water (Tortajada, 2006a, this volume). Dhaka is desperately short of finance and the capacity to provide the necessary water infrastructure to its inhabitants (Haq, 2006, this volume). Jakarta struggles with capacity shortcomings and institutional problems, which have not been eased by the privatization of the water sector in the 1990s (Lanti, 2006, this volume). Istanbul shows one path towards sustainable water management because the city has been subjected to massive investment in its water infrastructure and related environmental fields, as a consequence of strong political commitment to solving the city's water crisis of the early 1990s (Altinbilek, 2006, this volume). Johannesburg is in many ways on a similar path (Turton *et al.* 2006, this volume).

There is no question that the expansion of megacities implies the requirement of the provision of a water infrastructure to an extremely fast growing system (Braga, *et al.*, 2006, this volume; Lundqvist *et al.*, 2005). Whereas this is obviously the largest construction challenge and undertaking that the mankind has ever faced, this is not by far the whole picture (cf. Varis, 2006, this volume; Chapagain & Hoekstra, 2004), namely:

- The water issue in megacities is much more than a question of infrastructure (Varis, 2006, this volume).

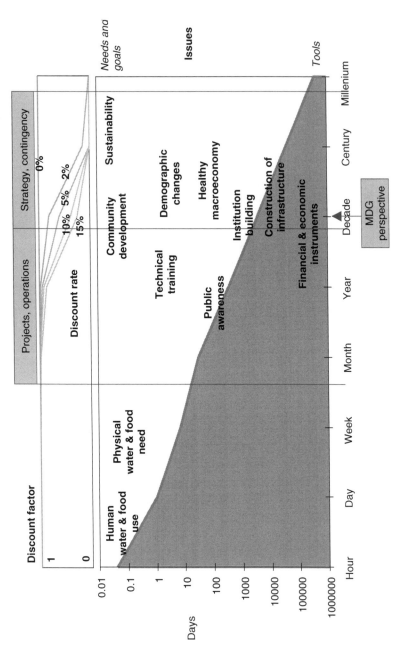

Figure 2. Sustainable development of megacities needs a long-term perspective, sound strategies and political commitment that go far beyond financial rationality or Millennium Development Goal perspective. *Source:* adapted from Varis & Somlyódy (1997).

- Water is not the only aspect of infrastructure development and provision in megacities: transportation, energy, housing, real estate, etc. planning should be developed and managed in an integrated way.
- Provision of water for the various sectors has the highest priority, and rightly so. But unless the challenges of 'water-after-use' is being attended, public health problems and serious economic repercussions from deteriorating water and environmental quality will increase and make sustainability of cities a mirage.
- The water footprint of a megacity goes far beyond the city limits. The world's megacities already import an amount of food comparable to the total international food trade. A bulk of this food has been produced using irrigation. They also import massive amounts of other resources such as energy, metals and fibre products etc., which are produced with huge interruptions to the hydrologic system, both qualitative and quantitative.
- The sources of many urban problems are rooted in the inadequate development of rural areas, which creates a high rural push type of migration pressure.
- A sustainable megacity needs a flourishing economy, strong social and environmental policies, a powerful and just governance system and adequate mechanisms for public participation. All these aspects are intertwined with the various roles of water in megacity development, and the water sector is often an important contributor to all these.
- Sustainable development of megacities needs a long-term perspective, sound strategies and political commitment that goes far beyond financial rationality or the Millennium Development Goal perspective.
- Equally important is a massive investment in infrastructure, including the water infrastructure, as the case studies have clearly shown.
- The IWRM concept should recognize the crucial role of large urban areas in which market links are even global and jurisdictions constitute a complex entity (many megacities are semi-autonomous areas). In this framework, talking about basin scale water management is insufficient. Integration is crucial as was pointed out by Tortajada (2006a,b, this volume) and Braga *et al.*, 2006, this volume. In the megacity context, issues such as those elaborated by Varis (2006, this volume; particularly Table 5) are rudimentary, albeit absent in the present IWRM definitions.
- Megacities are dramatic cases of urbanization and water related challenges, but it must be recognized that most megacities are national or even regional economical and political centres which gives them tools to approach their problems, which are superior to most medium sized cities.

References

Ahmad, E. & Kamal, M. M. (2004) Water management in Dhaka city, in: M. Q. Hassan (Ed.) *Water Resources Management and Development in Dhaka City*, pp. 33–38 (Dhaka: Goethe-Institut).

Abderrahman, W.A. (2006) Water Management in ArRiyadh, *International Journal of Water Resources Development*, 22(2), pp. 277–289.

Altinbilek, D. (2006) Water management in Istanbul, *International Journal of Water Resources Development*, 22(2), pp. 241–253.

Basson, M. S., van Niekerk, P. H. & van Rooyen, J. A. (1997) *Overview of Water Resources Availability and Utilisation in South Africa* (Pretoria: Department of Water Affairs and Forestry).

Braga, B. P. F., Porto, M. F. A. & Silva, R. T. (2006) Water management in Metropolitan São Paulo, *International Journal of Water Resources Development*, 22(2), pp. 337–352.

Biswas, A., Lundqvist, J., Tortajada, C. & Varis, O. (2004) Water management for megacities, *Stockholm Water Front*, 2, pp. 12–13.

Chapagain, A. K. & Hoekstra, A. Y. (2004) Water footprints of nations, Value of Water Research Report Series No. 16 (Delft: UNESCO-IHE).

Eroglu, V., Sarikaya, H. Z. & Aydin, A. F. (2001) Planning of wastewater treatment and disposal systems of Istanbul metropolitan area, *Water Science and Technology*, 44(23), pp. 31–38.

Haq, K.A. (2006) Water management in Dhaka, *International Journal of Water Resources Development*, 22(2), pp. 291–311.

IEA (2003) *Key World Energy Statistics 2003* (Paris: OECD/International Energy Agency).

INEGI (2000) *Statistics on the Metropolitan Area of Mexico City* (Mexico: National Institute for Statistics, Geography and Informatics).

Lanti, A. (2006) A regulatory approach to the Jakarta Water Supply concession contracts, *International Journal of Water Resources Development*, 22(2), pp. 255–276.

Lundqvist, J., Appasamy, P. & Prakash, N. (2003) Dimensions and approaches for third world city water security, *Philosophical Transactions of the Royal Society B: Biological Sciences*, 358, pp. 1985–1996.

Lundqvist, J., Tortajada, C., Varis, O. & Biswas, A. (2005) Water management in megacities, *Ambio*, 34, pp. 269–270.

McCartney, P. (2004) The case of Jakarta, Indonesia, in: *UN-Habitat Global Report on Human Settlements, The Challenge of Slums, Part IV: Summary of City Case Studies* (London: UN-HABITAT and Earthscan).

Segal, D. (2004) *Singapore's Water Trade with Malaysia and Alternatives*. John F. Kennedy School of Government (Boston: Harvard University).

Standl, H. (2003) Trinkwasser für die Mecacity Istanbul, *Geographische Rundschau*, 55(4), pp. 10–16.

Tortajada, C. (2003) Water management for a megacity: Mexico City Metropolitan Area, *Ambio*, 32, pp. 124–129.

Tortajada, C. (2006a) Water Management in Singapore, *International Journal of Water Resources Development*, 22(2), pp. 227–240.

Tortajada, C. (2006b) Water Management in Mexico City Metropolitan Area, *International Journal of Water Resources Development*, 22(2), pp. 353–376.

Turton, A., Schultz, C., Buckle, H., Kgomongoe, M., Malungani, T. & Drackner, M. (2006) Gold, scorched earth and water: the hydropolitics of Johannesburg, *International Journal of Water Resources Development*, 22(2), pp. 313–335.

UN (1989) *Population growth and policies in mega-cities: Jakarta* Population Policy Paper 18 (New York: United Nations).

UN (2002) *World Urbanization Prospects: The 2001 Revision* (New York: United Nations).

UNDP (2004) *Arctic Human Development Report* (New York: United Nations Development Programme).

UN HABITAT (2003a) *Slums of the World: The Face of Urban Poverty in the New Millennium?* (Nairobi: United Nations Human Settlements Programme).

UN HABITAT (2003b) *Water and Sanitation in the World's Cities: Local Action for Global Goals* (London United Nations Human Settlements Programme / EARTHSCAN).

UN HABITAT (2004) *The State of the World's Cities 2004/2005* (London: United Nations Human Settlements Programme/Earthscan).

Varis, O. & Somlyódy, L. (1997) Global urbanization and urban water: can sustainability be afforded?, *Water Science and Technology*, 35(9), pp. 21–32.

Zahid, A., Hossain, A., Uddin, M. E. & Deeba, F. (2004) Groundwater level declining trend in Dhaka city aquifer, in: M. Q. Hassan (Ed.) *Water Resources Management and Development in Dhaka City*, pp. 17–3? (Goethe-Institut: Dhaka).

INDEX

NOTE: Tables and figures are denoted by page numbers in italics and bold respectively

For Product Safety Concerns and Information please contact our EU
representative GPSR@taylorandfrancis.com Taylor & Francis Verlag GmbH,
Kaufingerstraße 24, 80331 München, Germany

Printed and bound by CPI Group (UK) Ltd, Croydon, CR0 4YY
02/05/2025
01859507-0001